T0252936

Meteorological Measurements and Instrumentation

Advancing Weather and Climate Science Series

Series Author:
R. Giles Harrison
Department of Meteorology, University of Reading, UK

Other titles in the series:

Mesoscale Meteorology in Midlatitudes
Paul Markowski and Yvette Richardson, Pennsylvania State University, USA
Published: February 2010
ISBN: 978-0-470-74213-6

Thermal Physics of the Atmosphere
Maarten H.P. Ambaum, University of Reading, UK
Published: April 2010
ISBN: 978-0-470-74515-1

The Atmosphere and Ocean: A Physical Introduction, 3rd Edition
Neil C. Wells, Southampton University, UK
Published: November 2011
ISBN: 978-0-470-69469-5

Time-Series Analysis in Meteorology and Climatology: An Introduction
Claude Duchon, University of Oklahoma, USA and
Robert Hale, Colorado State University, USA
Published: January 2012
ISBN: 978-0-470-97199-4

Operational Weather Forecasting
Peter Inness, University of Reading, UK and
Steve Dorling, University of East Anglia, UK
Published: December 2012
ISBN: 978-0-470-71159-0

Meteorological Measurements and Instrumentation

R. Giles Harrison
Department of Meteorology, University of Reading, UK

WILEY Blackwell

This edition first published 2015 © 2015 by John Wiley & Sons, Ltd

Registered office: John Wiley & Sons, Ltd, The Atrium, Southern Gate, Chichester, West Sussex, PO19 8SQ, UK

Editorial offices: 9600 Garsington Road, Oxford, OX4 2DQ, UK
The Atrium, Southern Gate, Chichester, West Sussex, PO19 8SQ, UK
2121 State Avenue, Ames, Iowa 50014-8300, USA

For details of our global editorial offices, for customer services and for information about how to apply for permission to reuse the copyright material in this book please see our website at www.wiley.com/wiley-blackwell.

The right of the author to be identified as the author of this work has been asserted in accordance with the UK Copyright, Designs and Patents Act 1988.

All rights reserved. No part of this publication may be reproduced, stored in a retrieval system, or transmitted, in any form or by any means, electronic, mechanical, photocopying, recording or otherwise, except as permitted by the UK Copyright, Designs and Patents Act 1988, without the prior permission of the publisher.

Designations used by companies to distinguish their products are often claimed as trademarks. All brand names and product names used in this book are trade names, service marks, trademarks or registered trademarks of their respective owners. The publisher is not associated with any product or vendor mentioned in this book.

Limit of Liability/Disclaimer of Warranty: While the publisher and author(s) have used their best efforts in preparing this book, they make no representations or warranties with respect to the accuracy or completeness of the contents of this book and specifically disclaim any implied warranties of merchantability or fitness for a particular purpose. It is sold on the understanding that the publisher is not engaged in rendering professional services and neither the publisher nor the author shall be liable for damages arising herefrom. If professional advice or other expert assistance is required, the services of a competent professional should be sought.

Library of Congress Cataloging-in-Publication Data applied for.

ISBN: 9781118745809

A catalogue record for this book is available from the British Library.

Wiley also publishes its books in a variety of electronic formats. Some content that appears in print may not be available in electronic books.

Set in 10/12pt Palatino by Aptara Inc., New Delhi, India

1 2015

Contents

Series Foreword

Advancing Weather and Climate Science

Meteorology is a rapidly moving science. New developments in weather forecasting, climate science and observing techniques are happening all the time, as shown by the wealth of papers published in the various meteorological journals. Often these developments take many years to make it into academic textbooks, by which time the science itself has moved on. At the same time, the underpinning principles of atmospheric science are well understood but could be brought up to date in the light of the ever increasing volume of new and exciting observations and the underlying patterns of climate change that may affect so many aspects of weather and the climate system.

In this series, the Royal Meteorological Society, in conjunction with Wiley Blackwell, is aiming to bring together both the underpinning principles and new developments in the science into a unified set of books suitable for undergraduate and postgraduate study as well as being a useful resource for the professional meteorologist or Earth system scientist. New developments in weather and climate sciences will be described together with a comprehensive survey of the underpinning principles, thoroughly updated for the 21st century. The series will build into a comprehensive teaching resource for the growing number of courses in weather and climate science at undergraduate and postgraduate level.

Series Editors

Peter Inness
University of Reading, UK

John A. Knox
University of Georgia, USA

"To make a real, significant impact on science you need to invent a new instrument"

James Lovelock[i]

Preface

Instrumentation and measurements have been a fundamental part of the undergraduate and postgraduate education in the Meteorology Department at the University of Reading since its foundation in 1965. The material here substantially follows the practical and theoretical course in *Measurements and Instrumentation* I began giving to first-term Applied Meteorology MSc students at Reading in 1994. This is an introductory course and necessarily includes a broad range of instrumentation and principles rather than solely specialised research techniques. Students begin with different quantitative and descriptive experiences, and, to date, have had destinies across environmental industries, research and meteorological services worldwide. A basic background in physics, including thermodynamics, statistics, electronics and experimental science, is assumed. In return, the understanding gained from the wide range of measuring and signal-processing techniques encountered in meteorology offers useful preparation for work in other areas of science.

Physical science itself is fundamentally an experimental subject, aiming to refine theories for observed phenomena, and atmospheric science is no exception. The range of physical environmental data available continues to increase as ever more imaginative measurement and dissemination technologies develop. In contrast, some of the core instrumentation has evolved only slowly, in part perhaps due to the inheritance of meteorological measurement methods by climate science, with its need to maintain consistency of record. Consequently, as with some key civil engineering infrastructure, Victorian technologies remain surprisingly important, despite meteorology often having rapidly adopted pioneering innovations such as the barometer, the balloon, the telegraph, the radio, electronics and of course the computer. Even these early technologies, however, still offer a useful context within which to appreciate modern instrumentation.

The modern abundance of data in the climate-related sciences implies those physical scientists developing instruments are unlikely to be the same individuals who subsequently analyse the results. This is because the spectrum of activity in measurements is substantial: it extends from simple inheritance of data files for analysis, through the use of commercial instruments for specific 'data-gathering' requirements, to conventional experimental science which invents apparatus as needed for accurate

[i] Remark in a seminar in the Department of Meteorology at Reading in 1999.

measurements or to generate entirely new knowledge. This represents a key distinction in approach between measurements in which familiarity with experimental deficiencies is fundamental, and the convenient, but poorly informed collection of data files where responsibility for their quality is devolved or even completely unknown. This book is intended to help bridge these very different communities of 'measurer', 'gatherer' and 'inheritor', through describing the underlying measurement science.

Acknowledgements

It is impossible to list all the stimulating and constructive interactions with students and colleagues concerning measurements and instrumentation which have occurred so easily at Reading over a long time, but I am especially grateful to Maarten Ambaum, Karen Aplin, Janet Barlow, Alec Bennett, Stephen Burt, Christine Chiu, Ford Cropley, George Dugdale, the late David Grimes and Alan Ibbetson, Anthony Illingworth, Roger Knight, Graeme Marlton, James Milford, Keri Nicoll, Mike Pedder, Keith Shine, Robert Thompson and Curtis Wood for discussions, some of whom also provided images. Miriam Byrne, once my fellow PhD student at Imperial College, gave advice on the book's structure and Karen Aplin also helped immensely with proofreading. I am glad to acknowledge the patience and encouragement of the many technical staff I have worked with, and, in particular, Andrew Lomas, whose support has been outstanding in developing many of the techniques and instruments described, as well as helping with diagrams and photographs for this book. I am fortunate that my parents equipped me with *Man Must Measure* by Lancelot Hogben at an early age, and also in acquiring the inspiration of my father's notes from an upper air (radiosonde) training course at Hemsby in 1957.

Disclaimer

I should add that many different manufacturers' products appear in images in this book. This should not be interpreted as either endorsement or criticism; it is merely my personal selection from good fortune in visiting a wide range of interesting experiment and measurement sites over many years.

1

Introduction

The appearance of the sky and its relationship to the atmosphere's properties have, no doubt, always provoked curiosity, with early ideas on explaining its variations available from Aristotle. A defining change in the philosophy of atmospheric studies occurred in the seventeenth century, however, with the beginning of quantitative measurements, and the dawn of the instrumental age. Since then, elaborate devices to monitor and record changes in the elements have continued to develop, providing, along the way, measurements underpinning the instrumental record of past environmental changes, most notably in air temperature. This means that characterising and understanding early meteorological instruments are of much more than solely historical interest, as recovering past measurements, whilst recognising their limitations, can also have immediate geophysical relevance.

An important meteorological example is the reconstruction of past temperature variations from the miscellaneous thermometer records originally undertaken to satisfy personal curiosity. Ships' logbooks provide another example, in terms of geomagnetic field changes. Beyond the actual data produced in either case, this also provides a reminder that all measurements can have unforeseen applications well beyond their original motivation [1], either through a change of context in which the measurements are evaluated, or because other subsequently important information has unwittingly been included.[i] Such future scope is probably impossible to predict completely, but it can to some extent be allowed for by ensuring a full appreciation of the related measurement science through careful description of the construction, calibration and recording procedures for the instrumentation employed. The possible future legacy implied by taking this historical perspective adds further motivation for rigour in the modern science of atmospheric measurement.

This chapter briefly highlights some of the major historical landmarks in development of instrumentation science for meteorology, and concludes with an overview of the book's material.

[i] Consider, for example, the paper burn made by sunlight originally devised to determine the daily duration of sunshine. It is now appreciated that this provides a permanent, continuous record of the detailed state of the sky (see Section 9.8.1).

Meteorological Measurements and Instrumentation, First Edition. R. Giles Harrison.
© 2015 John Wiley & Sons, Ltd. Published 2015 by John Wiley & Sons, Ltd.
Companion website: www.wiley.com/go/harrison/meteorologicalinstruments

1.1 The instrumental age

Many of the early atmospheric measuring instruments were developed in Florence, due perhaps in part to the experimental physical science tradition inspired by Galileo, and availability of the necessary craftsmanship. This included early thermometers, such as the thermoscope produced during the late 1500s to determine changes in temperature. Following key instrument advances such as the invention of the barometer by Evangelista Torricelli in 1643 and an awareness of the need for standardisation of thermometers, modern quantitative study of the atmosphere can be considered to date from the mid-seventeenth century.

Early measurement networks followed from the availability of measuring technologies combined with the formation of learned scientific societies, which together provided the means to record and exchange information in a published form. Comparison of measurements required a system of standardisation, such as that achieved through common instrumentation, and in many cases, common exposure. For thermometers, an agreed temperature scale was necessary and the Celsius,[ii] Fahrenheit[iii] or Réaumur[iv] scales all originated in the eighteenth century [2]. The meteorological values were published as tables of readings, in many cases without any further processing, but which were sufficiently complete for analysis to be made later.

1.2 Measurements and the climate record

Early weather records can be found in 'weather diaries', which were usually kept by well-educated and well-resourced individuals able to purchase or construct scientific instruments such as barometers and thermometers. In some cases, these diaries contain considerable descriptive and quantitative geophysical data, such as those of temperature and rainfall measurements (Figure 1.1).

Such early data sources are important because of the reference information they provide for the study of climate change, and they therefore remain of scientific value many centuries later. This is particularly true of the disparate thermometer measurements made in southern England from the 1600s, which, although made originally by individuals in an uncoordinated way, now provide an important climate data resource. The temperature readings were cross-checked and compiled[v] in the 1950s, drawing on knowledge of the different instruments used and understanding of their exposures [4]. This important synthesis generated a long series of temperature data for an area conveniently described as 'Central England', amounting to an approximately triangular region bounded by Bristol, Manchester and London.

[ii] Anders Celsius (1701–1744), professor of Astronomy at Uppsala, proposed his temperature scale in 1742. It originally used the melting and boiling points of ice and water as fixed points, reversed from the modern use, giving temperatures of 100 and 0 for freezing and boiling points respectively.

[iii] Daniel Fahrenheit (1686–1736) used the extremes of temperatures then available, producing a scale in 1724 with the melting point of ice at 32°F and the boiling point of water at 212°F. A Fahrenheit temperature F can be converted to a (modern) Celsius temperature C by $C = (5/9)(F-32)$.

[iv] Réaumur (Réné Antoine Ferchault) (1683–1757) used a scale with 0, the melting point of ice, and 80 the boiling point of diluted alcohol (78.3°C).

[v] This was compiled by the climatologist Gordon Manley (1902–1980) and first published (covering 1698 to 1952) in 1953. An updated and extended version (for 1659 to 1973) was published in 1974.

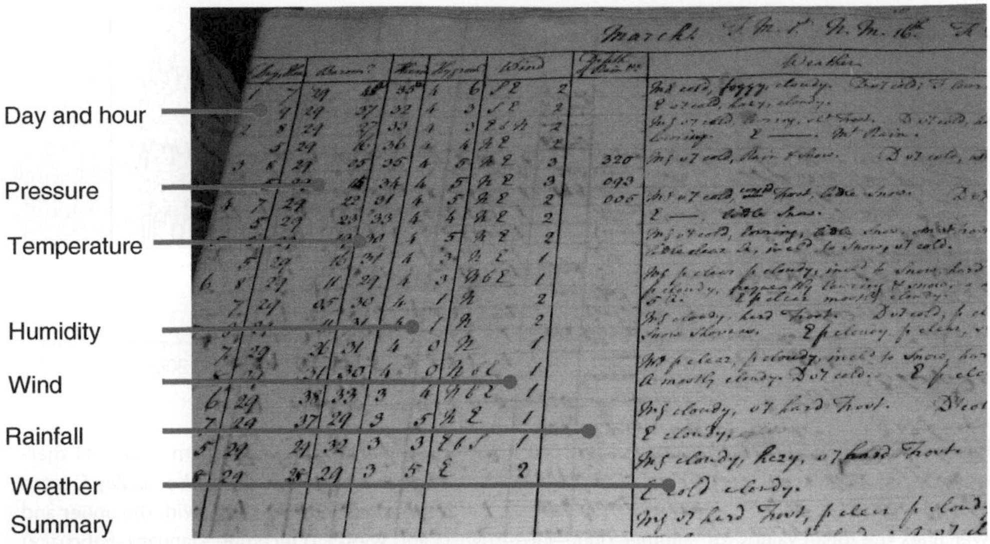

Figure 1.1 Example page from a weather diary (kept by an apothecary and surgeon, Thomas Hughes at Stroud, Gloucestershire, between 1771 and 1813), in which daily measurements of air pressure, temperature, humidity, rainfall and weather were recorded. As well as quantitative weather information, this particular diary includes other geophysical information, such as timings of earthquakes and even occurrence of the aurora borealis, an indirect measure of solar activity [3]. (Reproduced from Reference 3 with permission of The Met Office.)

The Central England measurements form the longest *continuous* set of monthly instrumental atmospheric temperatures available anywhere in the world, beginning in January 1659. (Daily values are also available, beginning in 1772; see Reference 5.) Figure 1.2 shows minimum, maximum and mean annual temperatures of the monthly Central England Temperature (CET) series.

1.3 Clouds and rainfall

In the nineteenth century, classification, quantification and taxonomy became an important aspect of many sciences, particularly in the life sciences and geology, so it was natural for similar approaches to be extended to meteorology. The classification of clouds[vi] was one early aspect, and the compilation of rainfall data also helped further develop the quantitative basis for environmental description. Major developments in meteorology continued in the mid-nineteenth century, following the foundation of the Meteorological Society in 1850, and the establishment of the early Met Office in 1854 under Admiral Fitzroy.[vii] The British Association for the Advancement of Science convened a Rainfall Committee, with G.J. Symons as secretary.

[vi] Luke Howard (1772–1864) established a classification system for clouds in 1802 (see Richard Hamblyn's *The Invention of Clouds*, published by Picador).

[vii] This is thoroughly discussed in *History of the Meteorological Office* by Malcolm Walker, published by Cambridge University Press.

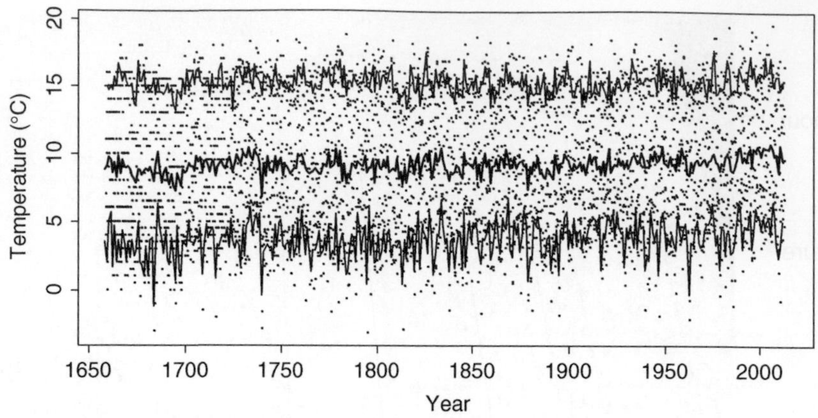

Figure 1.2 Monthly temperatures of 'Central England', originally constructed from historical thermometer records by Manley, and continued using updated modern measurements by the Hadley Centre of the UK Met Office [5]. The thick central line shows the annual mean temperature, with the upper and lower lines the mean values for summer (June–July–August) and winter (December–January–February) respectively (the degraded resolution of the early thermometers is also evident). (Reproduced from Reference 5 with permission of The Met Office.)

Compilation of historical rainfall data for the United Kingdom was a herculean undertaking, but, following adverts in many local newspapers leading to thousands of replies, Symons [6] did conclude in 1866 that 'there are not now very many records in private hands of which copies are not already obtained and classified.' The legacy of this work is the series of annual volumes of *Symons British Rainfall*. Further, a continuous series of monthly data [7] for England and Wales Precipitation (EWP) exists from 1766 (see figure 1.3).

1.4 Standardisation of air temperature measurements

Standardised exposure for air temperature measurements began in the nineteenth century [8], when meteorological instruments were becoming increasingly available

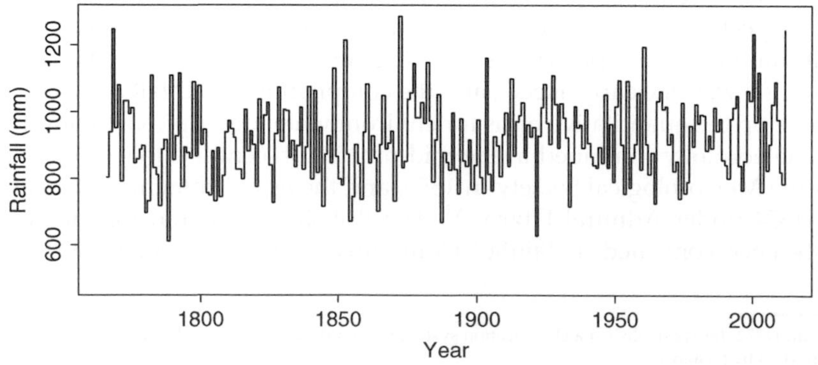

Figure 1.3 Annual rainfall for England and Wales. (Reproduced with permission of The Met Office.)

commercially.[viii] Early (1841) exposure of thermometers for air temperature measurement was through use of a Glaisher stand,[ix] a simple shading board which was rotated manually to prevent direct solar radiation reaching the thermometer [9]. The Glaisher stand's effectiveness depended on the diligence of the observer required to turn the stand after each reading. If the interval between readings became too long, direct sunlight could still reach the thermometer. The practical difficulty in manually turning the shade board yet retaining good ventilation was solved by Thomas Stevenson[x] in 1863, in the form of a double-louvered wooden box painted gloss white. This gave protection to thermometers from solar radiation in all directions, and ensured long wave radiation exchange was with the interior of the screen. The use of a double-louver increased the length of the air path through the screen, which brought the interior of the screen material closer to air temperature than alternatives of simple slits or mesh. In its original form, the Stevenson screen was a wooden box 15 inches high, 14.5 inches long and 7.5 inches wide. It had a solid roof with integral ventilator, and the thermometers were mounted horizontally 4 feet above the ground.

Many minor variants on the Stevenson screen were made. The Scottish physicist John Aitken investigated screen properties [10], noting much later [11] that nothing had been done to mitigate the effects of thermal inertia of Stevenson screens. Assessments of the Stevenson screen and the Glaisher stand were undertaken between 1868 and 1926 at a variety of locations [12] including a sustained 40 years of comparison at Camden Square [13]. The Glaisher stand thermometer was shown to read warmer in the summer months, by a maximum of 3.3°C. Many refinements were made to the Stevenson style screens from the original design, including a double roof and staggered boards across the base to exclude reflected radiation [14, 15], but it has remained largely unchanged since 1884. It was recommended for use by the Meteorological Society in 1873. (Properties of Stevenson screens are discussed in Chapter 5.)

1.5 Upper air measurements

The very first upper air measurements provided fundamental information on the atmosphere's structure. Kites were used for early soundings, such as for carrying thermometers aloft in 1749, and famously employed by Benjamin Franklin for thunderstorm studies in 1752. Manned ascents began with the hot air balloon of Montgolfier (1783), but lighter than air balloons provided suitable measurement platforms to obtain the atmospheric temperature profile to altitudes of several kilometres, notably by Gay-Lussac (1804). Instrumented balloons were later developed as an alternative exploratory tool, with which fundamental discoveries of atmospheric structure were made. For example, after over 200 instrument ascents by day and night, Leon Teisserenc de Bort in 1902 reported a temperature discontinuity at about 11 km, dividing the atmosphere into layers, identifying the troposphere and stratosphere.

[viii] For a survey of nineteenth century instruments, see: *A Treatise on Meteorological Instruments*, Negretti and Zambra, London, 1864, or Middleton's History of the Meteorological Instruments (Johns Hopkins press, 1969).

[ix] The Glaisher stand was originally designed by Sir George Airy (1801–1892), Astronomer Royal, for use at Greenwich Observatory.

[x] Thomas Stevenson (1818–87) was a civil engineer known for the design of many Scottish lighthouses, and father of the writer Robert Louis Stevenson.

Figure 1.4 The launch of the balloon *Mammoth*, carrying Coxwell and Glaisher, from Stafford Road gasworks, Wolverhampton, probably on 18 August 1862. (Reproduced with permission of John Wiley & Sons.)

1.5.1 *Manned balloon ascents*

At their earliest stage, manned ascents were voyages of discovery not without risk, indeed presenting mortal danger to those involved. Only a few scientific ascents [16] were made in the early years of the nineteenth century, but a surge of research flights occurred in the United Kingdom from about 1850, including a famous ascent of James Glaisher[xi] and Henry Coxwell from Wolverhampton. This particular flight probably reached about 7 km, although the aeronauts became unconscious and were unable to read their barometer. They were lucky to survive. There are good records of some of these balloon ascents (see Figure 1.4) and accounts of these flights provide not only quantitative information about the atmospheric conditions above the surface, but also insights into the difficulties faced by the aeronauts:

> *The weather on the day (Aug. 18, 1862) of the third ascent was favourable, and there was but little wind. All the instruments were fixed before leaving the earth. A height of more than 4 miles was attained, and the balloon remained in the air about two hours. When at its highest point there were no clouds between the balloon and the earth, and the streets of Birmingham were distinctly visible. The descent was effected at Solihull, 7 miles from Birmingham. On the earth the temperature of the air was 67.8°F, and that of the dew-point 54.6°F; and they steadily decreased to 39.5°F and 22.2°F respectively at 11,500 feet. The balloon was then made to descend to the height of about 3000 feet, when both increased to 56.0°F and 47.5°F respectively. On throwing out ballast the balloon rose again, and the temperature declined pretty steadily to 24.0°F, and that of the dew-point to −10.0°F at the height of 23,000 feet. During this ascent Mr Glaisher's hands became quite blue, and he experienced a qualmish sensation in the brain and stomach, resembling the approach of sea-sickness; but no further inconvenience, besides such as resulted from the cold and the difficulty of breathing, was experienced. This feeling of sickness never occurred again to Mr Glaisher in any subsequent ascent. (Encyclopaedia Britannica, 1902)*

[xi] James Glaisher (1809–1903) was a scientific assistant (and later Superintendent) at Greenwich Observatory, and one of the founders of the Meteorological Society in 1850 (subsequently the Royal Meteorological Society).

Table 1.1 Some significant early scientific balloon ascents in Europe (modified from Reference 17)

Investigator	Launch details	Height (m)	Met data: Temperature	Met data: Humidity	Other measured quantities or remarks
Robertson and Lhoest	1803 (18 July) Hamburg	7000			Atmospheric electricity
Gay-Lussac and Biot	1804 (24 August; 16 September), Paris	7015	√		Geomagnetism
Barral and Bixio	1850 (29 June; 27 July)	7050	√		
Coxwell and Glaisher	1862 (18 August), Wolverhampton	6900	√	√	
Tuma	1892, 1894 (22 September); 7 flights 1894 to 1898, Salzburg	3000	√	√	Atmospheric electricity
Le Cadet	1893 (1 and 9 August), Meudon-Valhermay, Paris	2520			Atmospheric electricity
Börnstein	1893, Berlin				Atmospheric electricity
Hess	1912 (7 August), Aussig	5350			Discovery of cosmic rays

Some early balloon explorations in the nineteenth century are summarised in Table 1.1, which show a steady increase in use of this measuring platform into the early twentieth century.

1.5.2 *Self-reporting upper air instruments*

Kite-carried instruments continued to be used for research at the end of the nineteenth century and into the early twentieth century, such as by Napier Shaw and W.H. Dines [18]. These carried early recording instruments, or *meteorographs*, which were highly technically innovative. In special configurations and at suitable sites, kite systems could reach up to 7 km [19]. The meteorograph developed by Dines recorded data mechanically by making indelible marks on a metal plate, plotting temperature and humidity against pressure. Later devices employing a rotating drum for recording data were carried on aircraft in the 1920s [20], around which time the use of kites for 'scientific aeronautics' largely ceased (see also Section 8.5), as aircraft platforms became more available. The use of related mechanical recording devices for atmospheric measurements on aircraft was pioneered by G.M.B. Dobson,[xii] and first implemented on military flights from Upavon in Wiltshire during 1916 [21].

The development of small shortwave radio transmitters permitted the information obtained to be sent instantaneously to a distant observer by radio telemetry, leading to the *radiometeograph*. Demonstration of this technology in the late 1920s

[xii] Gordon Miller Bourne Dobson (1889–1976) was an atmospheric and experimental physicist who became a professor at Oxford in 1945, and after whom a unit of ozone amount is named.

was led by P. Idrac and R. Bureau [22, 23], who showed short wave radio transmitters and a pulse-based method could signal measured temperature and pressure values, although the first radiosonde providing data to a meteorological service was launched by P. Molchanov from Pavlovsk on January 30 1930. Improvements in radio systems, data transfer and batteries led to commercial designs of radiosonde becoming available from about 1936 [24] (see also Section 11.1.)

1.6 Scope and structure

This book is intended to provide background and introductory material on instrumentation science as applied and required for meteorological measurements. It is not, however, in any sense a guide to observing practices or conventions, which are considered more thoroughly elsewhere.[xiii] Rather it considers aspects of instrument and measurement theory (Chapter 2), the electronics required for signal conditioning (Chapter 3) and digital data acquisition and logging (Chapter 4). A range of common instruments and sensors are explored (Chapters 5–10). In Chapter 11, the preceding material is drawn on to describe the combination of sensors, signal processing and data transfer required for radiosonde measurements, and Chapter 12 gives examples of some of the processing techniques used for analysing environmental data. The two brief appendices provide further information, and, in Appendix A, a summary of how a paper on new developments in instrumentation can be written is given.

[xiii] See, for example, *The Weather Observer's Handbook* by Stephen Burt (Cambridge University Press).

2

Principles of Measurement and Instrumentation

A measuring instrument or technique is a fundamental requirement in determining the magnitude of a physical environmental parameter such as temperature, wind speed or pressure. This requires that sensing technology is able to respond the quantity to be measured, a method of quantifying the response, and in some situations, a recording technique for the quantified response. Closely related is the need to evaluate how well the quantity has been measured, to provide an assessment of the associated uncertainty.

2.1 Instruments and measurement systems

An instrument in this context is a physical device or system, used to measure or monitor something. It may take the form of a hand-held unpowered device, such as a liquid-in-glass thermometer, or it may be physically substantial and require hardware and software to obtain and process the results (e.g. a radiosonde receiver and processor), in which case it would be more appropriately regarded as a measurement system. Figure 2.1 summarises the elements likely to be present in an instrument or measurements system designed to respond to a particular parameter.

The *sensor* responds to the specific parameter measured. (In some cases, energy exchange by a *transducer* may also be required, to convert the sensor's response into something which can be more conveniently measured.) An *amplifier* is used to increase the magnitude of the changes produced by the sensor. Amplifiers operate on a variety of principles, for example mechanical, chemical, optical or electronic. As well as increasing the *signal*, an amplifier usually increases other random variations (*noise*) present as well. The *meter* provides the final readout in terms of a magnitude, and can be digital or analogue. A recording device of some form may be attached to the meter, such as a chart recorder, a computerised logging system or, more simply, an observer with a notebook.

Key terms commonly used to describe the principal aspects of a measurement system are listed in Table 2.1.

Meteorological Measurements and Instrumentation, First Edition. R. Giles Harrison.
© 2015 John Wiley & Sons, Ltd. Published 2015 by John Wiley & Sons, Ltd.
Companion website: www.wiley.com/go/harrison/meteorologicalinstruments

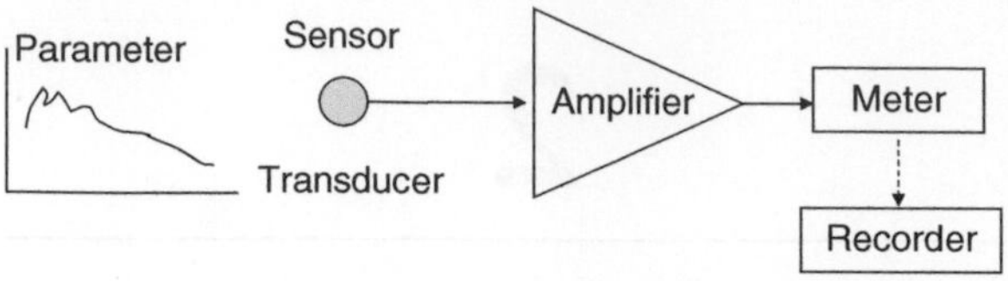

Figure 2.1 Elements of a typical instrument or measuring system intended to quantify variations in the parameter to be sensed. Some measurement systems also include the ability to record the variations obtained. (See Table 2.1 for definitions of the terms used.)

2.1.1 Instrument response characterisation

The response of an instrument to the parameter sensed can be characterised in several ways, concerning how large the actual response is, how quickly the instrument responds, how reliable the readings obtained are and over what range of the input parameter the instrument functions reliably. All these considerations serve to describe how the instrument response can be interpreted to determine quantitative changes in the parameter sensed.

The most fundamental use of any sensor or instrument is as a detector to determine the presence or absence of a change in a parameter and the *resolution* (sometimes also known as the discrimination) describes this smallest detectable change. This can be found experimentally by decreasing the parameter sensed until the instrument no longer responds. The quantitative response to the change is described by the *sensitivity*, which is the ratio of the response in the instrument to a unit change in the parameter sensed. For an instrument with a linear response, the sensitivity will be the same no matter how large or small the value of the parameter, yielding a straight line relationship between variations in the parameter and the instrument's response

Table 2.1 Descriptions of parts of an instrument

Term	Explanation
Parameter	The variable physical quantity to be measured, such as pressure, temperature, speed, time.
Sensor	A device which responds directly in some way to changes in the parameter to be measured. For example, the rotating cups on a cup anemometer.
Transducer	An energy transfer device to convert the response of a sensor into another quantity, which can be more conveniently measured (e.g. to an electrical signal) or recorded. In some instruments, such as a photovoltaic light meter, the functions of sensor and transducer are combined.
Amplifier	This magnifies a small change, for example turning a small voltage change into a larger voltage change; amplifiers are, however, not necessarily electronic.
Gain	The ratio of the output signal amplitude to the input signal amplitude of an amplifier stage.
Meter	This measures and displays the output from a transducer or sensor (meters are often electrical, in that they display a voltage, but may also be mechanical).
Recorder	A device employing a retrieval method (paper, film, electronic etc.) by which a series of successive meter readings can be preserved.

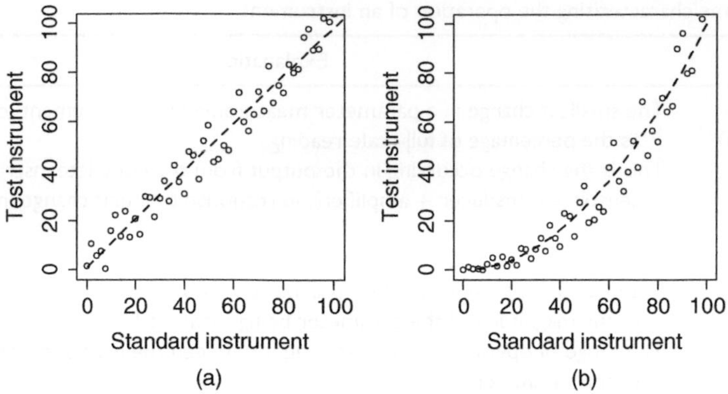

Figure 2.2 Points represent the responses of a standard instrument and an instrument being characterised to changes in a sensed parameter for (a) a linear instrument and (b) a test instrument with a non-linear response. Dashed lines represent the derived response characteristic of the instrument being characterised.

(Figure 2.2a). In this case, for the output plotted as a function of the input, the gradient of the line described would be the instrument's sensitivity. If the sensitivity varies with the parameter sensed, the instrument has a non-linear response (Figure 2.2b). In the non-linear case, it may still be possible to characterise the response by a mathematical function, such as by several terms of a polynomial function or, alternatively, the linear responses at several key values of the input parameter can be used separately.

There are other factors which can influence an instrument's response to variations in the parameter sensed. If the instrument is unable to measure over the full range of the possible values of the parameter, it is said to have insufficient *dynamic range*. The unfortunate consequence of the parameter exceeding the dynamic range is that the instrument can only return its maximum (or minimum) value, and there will be no further variation measured even if there are continued variations above this maximum (Figure 2.3a). The instrument is then said to be saturated, and the only

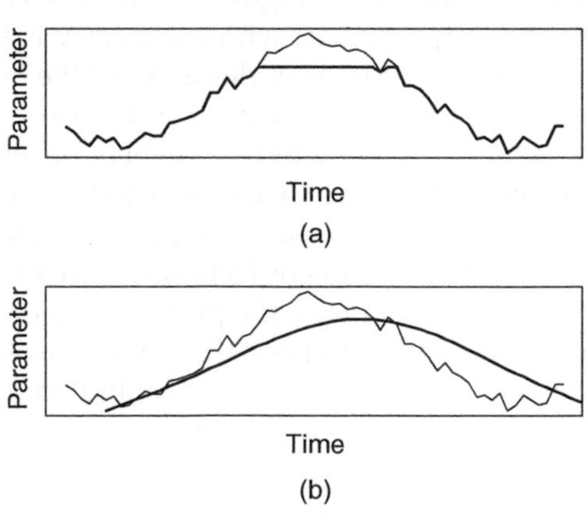

Figure 2.3 Variations in a parameter (thin line) sampled by an instrument (thick line) with (a) insufficient dynamic range and (b) limited time response.

Table 2.2 Terms characterising the operation of an instrument

Term	Explanation
Resolution (or discrimination)	The smallest change in a parameter measurable by an instrument, often quoted as the percentage of full-scale reading.
Sensitivity	This is the change occurring in the output from the complete instrument (*i.e.* sensor + transducer + amplifier), in response to a unit change in the parameter sensed.
Linearity	If the responses of the sensor and other components (such as the transducer and amplifier) are linear, then the instrument's sensitivity will be independent of the magnitude of the parameter being measured.
Dynamic range	The range of operation of an instrument between the least and greatest value of the parameter.
Time response	A measure of how long an instrument takes to respond by a required amount to a sudden change in the parameter sensed.
Characteristic	The relationship between the sensed parameter and an instrument's measurements. The response characteristic is especially important for non-linear instruments.

information it provides is that the parameter is greater than or equal to the instrument's maximum value, with an indication of for how long the saturation condition persists.

An instrument may also not be able to follow rapid variations in the parameter sensed, because of a limited *time response*. The consequences are firstly that rapid changes in the parameter become smoothed out by the instrument and so are not determined, and secondly that there may also be some delay in the changes being registered (Figure 2.3b). Table 2.2 summarises terms used in describing an instrument's response.

2.1.2 Measurement quality

If the instrument comparison shown in Figure 2.2 is between an instrument of unknown response and another instrument whose accuracy is known, the comparison constitutes a calibration. Calibration provides a test of the instrument's response against known values of the parameter sensed, or at least values of the parameter which are themselves known to an acknowledged level of uncertainty.

The successive comparison of a poorer instrument with a better instrument ultimately leads to the need for a definitive instrument or calibration standard. In general, such standards, reference instruments or techniques are maintained by national standards laboratories. These are essentially the primary standards, from which secondary standard instruments are calibrated and distributed to other laboratories. The secondary standard instruments, and instruments calibrated against them in turn, are then in principle traceable to the national standards. There are also 'absolute' instruments which can be self-calibrating, through permitting a comparison with other physical quantities independently defined. The cavity radiometer (Section 9.6.1) provides an example in which the effect of radiative heating is compared with well-determined electrical heating, and the pitot tube anemometer (Section 8.2.2) determines wind speed from a pressure difference. In these two cases, the calibration

Figure 2.4 A viewpoint perhaps not wholly confined to experimental uncertainty, expressed in Lower Goat Lane, Norwich.

does not require known amounts of radiation or fixed wind speed, but instead reliable measurements of electrical power (hence requiring well-characterised current and voltage instruments) and pressure (by implication ultimately needing well-characterised force and length determinations) respectively.

Calibration of an instrument will reveal the uncertainties (or errors)[i] associated with its measurements. In general, the uncertainties arise from two sources, *systematic* effects (or biases) which are endemic to the specific instrument or the basis on which it operates and *random* effects due to sampling. The general principle of reducing uncertainties (see also Figure 2.4) is a key goal of a well-designed experiment. In general, systematic effects can be reduced by careful calibration, sometimes also known as bias correction; the random effects can be reduced by averaging, with the effectiveness depending on how many samples are available.

An instrument is regarded as precise if, every time it measures the same value of a quantity, it returns a very similar reading within the range expected solely by statistical sampling considerations. Of course the consistency of reading says nothing about whether the instrument is also accurate, as it may be very consistent in returning a value which differs substantially from the true value of the quantity concerned. Good precision is therefore a necessary but not sufficient condition for the instrument to be

[i] *Uncertainties* have also been commonly known as *errors* in physical science. Of the two, the term uncertainty seems increasingly preferable, as it alludes to the fundamental limits of knowledge rather than implying a mistake has been made which might be simply corrected.

Table 2.3 Terms characterising the quality of a measuring instrument's performance

Term	Explanation
Calibration (or standardisation)	The process of comparing one instrument (or standard values) with another instrument whose characteristics are known more accurately.
Systematic uncertainty	A consistent, repeated offset in a measurement, as a result of a fixed or regular discrepancy in the instrument response (e.g. a bent needle on an analogue meter).
Random uncertainty	Variations in measurements due to statistical fluctuations in either the quantity sensed, the internal operation of the instrument, or a combination of the two.
Precision	An instrument is precise if, in repeated trials, it is able to give the same output response for the same value of the input parameter. Precision is therefore determined by the systematic error in the instrument remaining constant (note that a precise instrument may still be inaccurate).
Accuracy	A measure of the overall uncertainty in the value of the parameter measured, when compared against an external standard, often quoted as a percentage of full-scale reading. Accuracy is determined by the combination of random and systematic errors in the instrument (note that a sensitive, high resolution instrument may still be inaccurate).

considered accurate; a calibration against a better instrument or reference standard is also required.

Instrument accuracy may be reported in one of two ways:

1. as *absolute* accuracy, for example, ±0.1°C for temperature. In this case, the implied uncertainty does not depend on the actual value of temperature recorded by the instrument;
2. as *proportional* accuracy, for example, ±5% for wind speed. In this case, the implied uncertainty *does* depend on the actual value of wind speed recorded by the instrument. Occasionally, however, accuracy may be quoted as a proportion of a *full-scale deflection*, and specified for a certain operating configuration of the instrument concerned.

Table 2.3 summarises terms used to describe the quality of the measurements made by an instrument or measurement system.

2.2 Instrument response time

A real sensor or instrument will always take a finite time to respond to a change in the parameter it is determining. This will limit the instrument's response to a sudden change, or its ability to follow fluctuations.

2.2.1 *Response to a step change*

A straightforward method to determine the response time of an instrument is to consider its measurements following an instantaneous (or step) change in the parameter being measured, for example a sudden change in temperature. If an ideal thermometer, initially at a temperature T_0, is suddenly placed in an air stream whose temperature is T_a, the thermometer's temperature T will change at a rate proportional to the temperature difference.[ii] Thus the rate of change of the thermometer's temperature T with time t is given by

$$\frac{dT}{dt} = -\frac{(T - T_a)}{\tau}, \tag{2.1}$$

where τ is a constant of proportionality, with dimensions of time. This describes a first-order response to a step change. Integrating, using the initial condition that $T = T_0$ at $t = 0$ gives the variation in temperature with time $T(t)$ as

$$T(t) = T_a + (T_0 - T_a)\exp\left(-\frac{t}{\tau}\right). \tag{2.2}$$

From this it is apparent that eventually, as $t \to \infty$, the exponential term will diminish and the thermometer's temperature $T \to T_a$. Equation 2.2 also shows that τ is the relevant timescale of the system, and it is known as the exponential response time. After a time τ, the thermometer will have registered ~63% of the total change $(T_0 - T_a)$, while after 3τ, it will have registered ~99% of the change (see Figure 2.5). Hence, if τ is small, the device will respond rapidly to a step change.

The exponential response time of a thermometer can therefore also be found from an experiment in which temperatures are measured following a thermal step change, such as removing a thermometer suddenly from warm water. Equation 2.2 implies

$$\ln[T(t) - T_a] = \ln(T_0 - T_a) - \frac{t}{\tau}, \tag{2.3}$$

hence a graph of $y = \ln[T(t) - T_a]$ versus $x = t$ should yield a straight line of the standard form $y = mx + c$, with gradient $m = -1/\tau$ and intercept $c = \ln(T_0 - T_a)$.

2.2.2 *Response to an oscillation*

Atmospheric temperatures commonly fluctuate, and isolated step changes are rare.[iii] Oscillatory temperature fluctuations can be represented mathematically using a periodic function, such as a sine wave. If a thermometer is placed in air in which the temperature T_a oscillates sinusoidally about a mean temperature T_m with amplitude A and frequency f, as

$$T_a(t) = T_m + A\sin(2\pi ft), \tag{2.4}$$

[ii] This example is related to *Newton's law of cooling* which says that the rate of heat loss is proportional to the temperature excess of the thermometer over its surroundings.

[iii] Step changes can be seen in data from balloon-carried temperature sensors rising rapidly, or as a result of phase changes (see Figure 5.6).

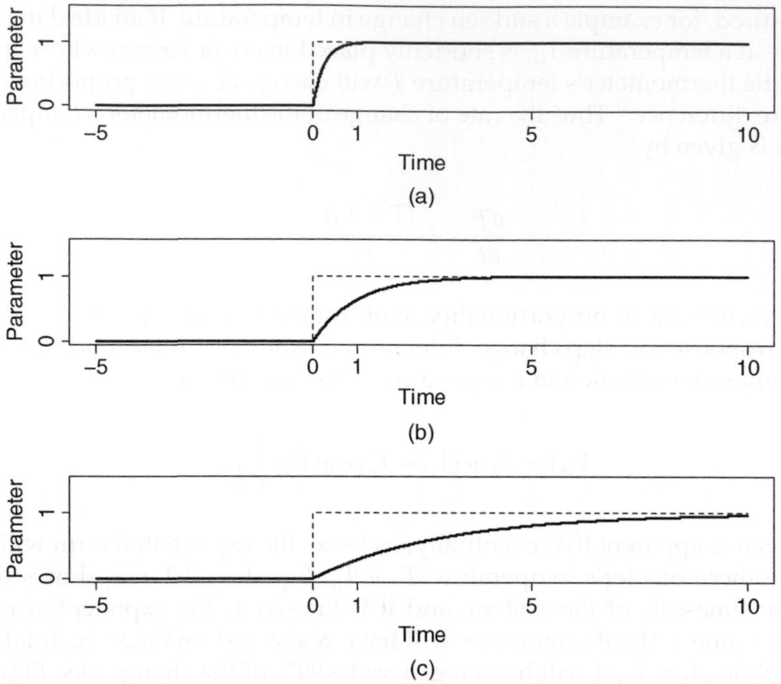

Figure 2.5 Response (solid line) to an instantaneous step change (dashed line) in the parameter sensed at $t = 0$, for an instrument with exponential response time τ (arbitrary units) of (a) 0.25, (b) 1 and (c) 3.

then the rate of change of temperature with time is

$$\frac{dT}{dt} = -\frac{1}{\tau}(T - T_a) = -\frac{1}{\tau}[T - T_m + A\sin(2\pi ft)]\,. \tag{2.5}$$

The time-dependent solution to Equation 2.5 is

$$T(t) = T_m + Be^{-t/\tau} + \frac{A}{\sqrt{1 + 4\pi^2 f^2 \tau^2}} \sin(2\pi ft - \varphi)\,, \tag{2.6}$$

where the phase angle $\phi = \tan^{-1}(2\pi f\tau)$ and the $Be^{-t/\tau}$ term diminishes with time following an initial transient to give a steady-state oscillatory solution (see Figure 2.7). Equation 2.6 shows that the thermometer will still indicate the correct mean temperature T_m, but that its response to the oscillation will be reduced in amplitude and shifted in phase so that its response lags the driving oscillation (see Figure 2.6).

Temperature fluctuations having timescales comparable with the thermometer response time τ are barely registered. For the thermometer amplitude response to exceed 50% of the real temperature fluctuations, the response time of thermometer must be less than the time scale of the fluctuations by a factor of at least four.

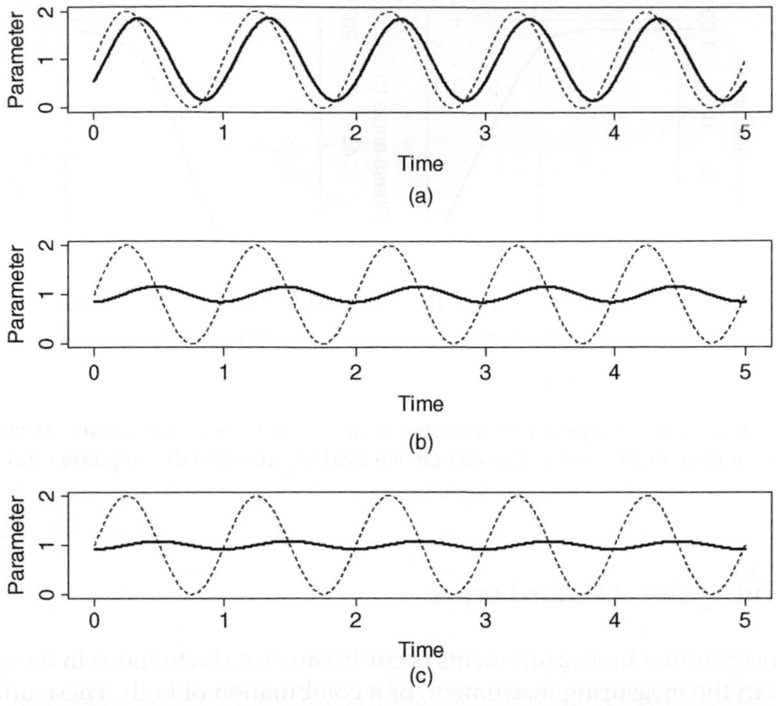

Figure 2.6 Response of an instrument to a fluctuating quantity (of oscillation frequency 1 unit, and oscillation amplitude 1 unit about a mean value of 1 unit) with an exponential time constant of (a) 0.1 units, (b) 1 unit and (c) 2 units.

The fractional amplitude reduction is given by the instrument *response ratio* R as

$$R = \frac{1}{\sqrt{1 + 4\pi^2 f^2 \tau^2}}. \tag{2.7}$$

An amplitude reduction to $R = \sqrt{1/2}$ is used to define a cut-off frequency f_c which also provides an estimate of the *bandwidth* of the system. The cut-off (or corner) frequency[iv] f_c is given by

$$f_c = \frac{1}{2\pi\tau}. \tag{2.8}$$

Thus the response time τ, as determined by a step change experiment, also characterises the frequency response of a first-order system. The effects on the fall-off in the amplitude response and increased phase lag with increasing frequency are shown in Figure 2.7, for an instrument with an exponential response time of $\tau = 1$ s.

[iv] f_c is also known as the '3 dB frequency', as, in decibels (dB), 3 dB represents a voltage reduction of $\sqrt{1/2}$ or a halving in power.

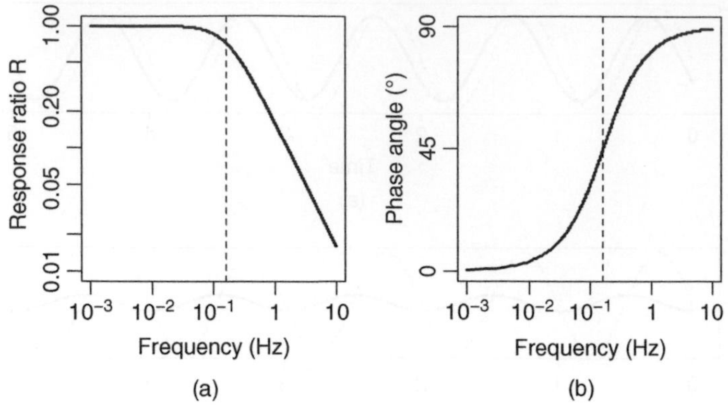

Figure 2.7 Variation with frequency of (a) response ratio and (b) phase lag, for an instrument with an exponential time constant $\tau = 1$ s. The vertical line marks f_c at which the response ratio becomes $\sqrt{1/2}$.

2.3 Deriving the standard error

Random uncertainties in measurements occur because of fluctuations in the quantity itself, noise in the measuring instrument, or a combination of both. These uncertainties are usually assumed to be normally distributed, so that the values x lie symmetrically around a mean value μ with a standard deviation σ, according to

$$f(x) = \frac{1}{\sigma\sqrt{2\pi}} \exp\left(-\frac{1}{2}\left[\frac{x-\mu}{\sigma}\right]^2\right), \tag{2.9}$$

which is normalised so that $\int f(x)dx = 1$. According to this distribution, (see Figure 2.8), 68% of values lie within $\pm 1\sigma$ of the mean μ, and 95% of the values lie within $\pm 1.96\sigma$.

The influence of the random fluctuations will be most apparent in any one measurement. This limits the ability of one measurement to give a good estimate of the quantity concerned. The effect of the random fluctuations can be reduced by repeated sampling and averaging, so that, in combination, their mean value provides a better estimate of the quantity sought. Not only does the set of measurements provide an improved estimate of the quantity sought, but it also provides an estimate of the spread in the measurements and a measure of the uncertainty – the standard error in the mean – associated with the mean value itself.

2.3.1 *Sample mean*

Consider a set of measurements of the same parameter X, represented by $X_1, X_2 \ldots X_i \ldots X_N$, where there are N measurements in total. These could represent a set of repeated, but independent, experimental measurements of a parameter

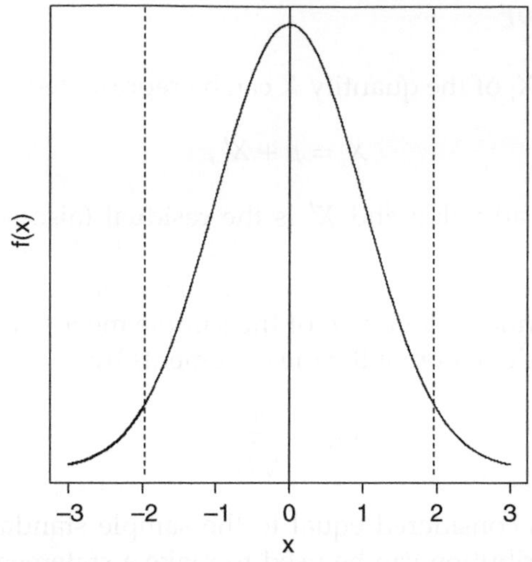

Figure 2.8 Normal distribution function $f(X)$, plotted for a mean $\overline{X}$ of zero and unit standard deviation σ. The dashed vertical lines enclose 95% of the values.

expected to have a constant value, such as the exponential response time of a thermometer.[v] From these values, the arithmetic mean $\overline{X}$ would be found as

$$\overline{X} = \frac{1}{N}(X_1 + X_2 \ldots + X_N) = \frac{1}{N}\sum_{i=1}^{N} X_i . \quad (2.10)$$

However, as $\overline{X}$ is only calculated from a finite 'sample' of all the possible measurements, it is often known as the sample mean. Generally, this will not be the same as the true mean value of the parameter μ, due to the effect of the limited sampling as X varies from one measurement to another. An important problem is, therefore, to estimate the accuracy of the sample mean, which depends on the variation (or dispersion) of the measurements around the true mean. This is usually estimated from the sample $(X_1 \ldots X_N)$ and sample mean $\overline{X}$ as the sample standard deviation, given by

$$s = \left\{ \frac{1}{(N-1)} \sum_{i=1}^{N} (X_i - \overline{X})^2 \right\}^{1/2} , \quad (2.11)$$

where X_i represents one individual sample, and the sample mean $\overline{X}$ has been previously calculated.[vi]

---

[v] If the measurements formed successive samples of a varying parameter at constant intervals of time at a fixed location, the data sequence is known as a *time series*. An example is the observation of turbulent fluctuations of near-surface temperatures using a fast-response thermometer.

[vi] If the measurements contain a large proportion of well-scattered values (outliers), the median may be a preferable measure rather than the mean, as it is less prone to the effect of outliers, particularly for small samples. There are corresponding 'robust' measures of spread, such as the separation between the upper and lower quartiles, known as the interquartile range (IQR). For a normally distributed variable, its standard deviation is IQR/1.349.

2.3.2 *Standard error*

A single measurement X_i of the quantity X can be represented by

$$X_i = \mu + X_i', \tag{2.12}$$

where μ is the true mean value and X_i' is the residual (also sometimes called the anomaly). If the residuals are independent, and the residuals are normally distributed about the true mean, the distribution of the sample means about the true mean will also be normal. The standard deviation of the sample means about the true mean is related to the standard deviation of the measurements by

$$\sigma_m = \frac{\sigma}{\sqrt{N}}. \tag{2.13}$$

If σ in Equation 2.13 is considered equal to the sample standard deviation s from Equation 2.11, this distribution can be used to make a statement of the accuracy of the measured mean value. This is because Equation 2.13 defines the useful quantity known as the standard error of the mean, σ_m. From Figure 2.8, there is approximately a 67% chance that μ lies within the range $\overline{X} \pm \sigma_m$, and the range corresponding to a 95% chance of including the true mean is given by $\overline{X} \pm 1.96\sigma_m$, which is a convention adopted for reporting experimental results. It is also clear that the standard error can be reduced if N (the number of measurements or experiments) is increased, but, because of the square root dependence, the initial improvement in standard error falls off quite rapidly with the number of determinations.

2.3.3 *Quoting results*

Reporting the probable error in a quantity is important because of the implications it has for calculations subsequently undertaken using that quantity or the broader interpretation of a finding, although it is rare to see numerical values characterised in this way outside of physical science (see Figure 2.9 for an exception). To a statistician, probable error has a rather precise meaning, expressing the probability that the true value of some parameter lies within the range $X \pm \delta X$, where X is the measured (or estimated) value of the parameter and δX is its probable error. In environmental science, the probable error tends to be used less formally, so that the statement

'measured value of temperature = $(T \pm \delta T)$°C'

would generally be taken to mean simply that there is a very high probability that the true temperature lies within the range of values given.

To take a specific example, suppose a measured temperature is recorded as 18.753°C and its error is estimated to be 0.125°C. It is realistic to say that the value of the measured temperature is (18.8 ± 0.1)°C. As uncertainty estimates are nearly always approximate, it is not usually justifiable to quote these to more than one or, at most, two significant figures. Approaches vary in rounding the final quoted value of a measurement, but the number of significant figures given should clearly be consistent with the uncertainty estimate, as there is little point in including many extra

Figure 2.9 Entrance hoarding at an Icelandic exhibition concerning the settlement of Reykjavik, which contains excavations dating to before 871±2. This date range was obtained from volcanic deposits and glacial records.

digits merely within the expected uncertainty. If the quantity is large or small, standard form can be used to simplify the reporting, but it is helpful in allowing comparisons to keep the same multiplier for both the measurement and the uncertainty, for example, 'droplet diameter = $(0.5 \pm 0.01)\times10^{-5}$ m'.

2.4 Calculations combining uncertainties

Once an uncertainty estimate has been made in a quantity measured, further calculations will often be needed using the value obtained, perhaps requiring it to be combined with another value or values also with uncertainties. Clearly, a calculation using two or more quantities each with different uncertainties will produce a new result which also has an uncertainty. In some cases, the uncertainty in one quantity will dominate over all the uncertainties in the others, and this may be regarded as the principal uncertainty. In general, depending on the exact calculations required, the uncertainties on the different quantities will have to be combined in some way.[vii] The uncertainties are said to *propagate* through the calculations.

2.4.1 *Sums and differences*

Suppose a quantity Z is calculated from the sum of two measured parameters A and B, so that $Z = A + B$. If, in a single pair of measurements, ε_A is the actual uncertainty on A and ε_B the actual uncertainty on B, then the actual uncertainty on Z is given by

$$\varepsilon_Z = \varepsilon_A + \varepsilon_B \,. \tag{2.14}$$

[vii] A simple method to estimate the spread in results from a calculation is to repeat the calculations with the term which gives the principal uncertainty perturbed to its greatest and least value.

To determine the accuracy of the result Z given the accuracy of the individual measurements from which it is found, the conventional approach is to consider the mean square error. From Equation 2.14, this is given by

$$\overline{\varepsilon_Z^2} = \overline{\varepsilon_A^2} + 2\overline{\varepsilon_A \varepsilon_B} + \overline{\varepsilon_B^2}, \tag{2.15}$$

where the over-bar means 'averaged over a very large number of independent measurements'. If the uncertainties on A and B are independent, which is generally the case, the correlation between their errors should be zero and hence the middle term on the right-hand side of this expression can be neglected. By assuming that squared probable errors can be combined in the same way as mean square errors, it follows that the probable error on Z is

$$\Delta Z = \sqrt{(\Delta A)^2 + (\Delta B)^2}, \tag{2.16}$$

where ΔA and ΔB are the probable errors on the individual measurements of A and B. The same principle, that the probable error for sums or differences can be estimated by taking the square root of the sum of the squares of the probable errors on the individual measured parameters, can be extended when Z is calculated from more than two measured parameters.

2.4.2 *Products and quotients*

If the quantity Z is found by multiplication, such as from

$$Z = AB, \tag{2.17}$$

when there is again uncertainty in each of A and B, the uncertainty in Z can be found by considering the uncertainties in A and B independently. Differentiating with respect to A, the variation in Z resulting from a small change in A is

$$\delta Z|_A = B\delta A = \frac{Z}{A}\delta A, \tag{2.18}$$

and that due to a small change in B is

$$\delta Z|_B = A\delta B = \frac{Z}{B}\delta B. \tag{2.19}$$

The combined fractional uncertainty from both terms is

$$\frac{\delta Z}{Z} = \sqrt{\left(\frac{\delta A}{A}\right)^2 + \left(\frac{\delta B}{B}\right)^2}. \tag{2.20}$$

The principle that the fractional error for products or quotients can be estimated by taking the square root of the sum of squared fractional errors can be extended when Z is calculated from more than two measured parameters.

Table 2.4 Summary of methods for estimating the resultant uncertainty for common combinations of variables *A* and *B*, each with individual uncertainties ΔA and ΔB

Functional relationship	Resultant uncertainty or fractional uncertainty in Z
$Z = A \pm B$	$\Delta Z = \sqrt{(\Delta A)^2 + (\Delta B)^2}$
$Z = AB$	$\left(\frac{\Delta Z}{Z}\right)^2 = \left(\frac{\Delta A}{A}\right)^2 + \left(\frac{\Delta B}{B}\right)^2$
$Z = \frac{A}{B}$	$\left(\frac{\Delta Z}{Z}\right)^2 = \left(\frac{\Delta A}{A}\right)^2 + \left(\frac{\Delta B}{B}\right)^2$
$Z = A^n$	$\left(\frac{\Delta Z}{Z}\right) = n\left(\frac{\Delta A}{A}\right)$

2.4.3 *Uncertainties from functions*

The effect of an uncertainty can be surprisingly large if the quantity concerned is raised to a power or is the input argument to another non-linear function. Consider calculating Z from a measured parameter A using a relationship of the form

$$Z = kA^n, \tag{2.21}$$

for k a known constant.[viii] The derived uncertainty in Z resulting from the uncertainty in A can be found by differentiating Equation 2.21 to give

$$\delta Z = nkA^{n-1}\delta A \equiv n\frac{Z}{A}\delta A, \tag{2.22}$$

where δA represents a small but finite uncertainty in A. The fractional uncertainty error in the derived Z is therefore

$$\frac{\delta Z}{Z} = n\frac{\delta A}{A}. \tag{2.23}$$

To combine the uncertainties for other functions when two or more measured parameters are involved, the above principles are applied to each parameter in turn, and then the results are combined.

A summary of the evaluation of the overall uncertainty for some common calculations is given in Table 2.4.

2.5 Calibration experiments

A common experiment is the comparison of one instrument with another, for calibration.[ix] This will yield a set of measurements from both instruments, which can be

[viii] This type of power law occurs in the Stefan–Boltzmann formula for emitted radiation, with $n = 4$. The fractional uncertainty in the derived radiative flux density is four times the fractional error in the absolute temperature.

[ix] It may not be necessary to actually perform such an experiment, as the variations needed in both instruments may be provided by atmospheric changes occurring naturally – see Section 4.4.1.

plotted against each other. If the instruments have linear responses, then a straight line may be drawn through the points. In drawing such a line, the intention is to provide a good fit to the data, and then to use the properties of the line – generally the gradient as it provides the sensitivity – to provide the calibration of the instrument under test. The line may be fitted in a variety of ways, by eye, using a transparent ruler, or by using a statistical computing package that is able to fit the line to meet some optimisation criteria. These criteria consider the difference between the line and the points, seeking, typically, to minimise the squares of the difference ('least squares'), or the absolute difference.

Figure 2.10 shows measurements obtained during characterisation of a sensitive current measuring device (a picoammeter), by measuring its output voltage when its input current is varied. The picoammeter has a linear response to input current, and Figure 2.10a shows the measurements obtained together with a best-fit line added using a statistical package. The implication of solely using points for the plotting is that the measurements are not subject to uncertainty, and therefore that any uncertainty only results from the methodology chosen in fitting the line. In reality, it is likely that both the input parameters and output parameters of the calibration will

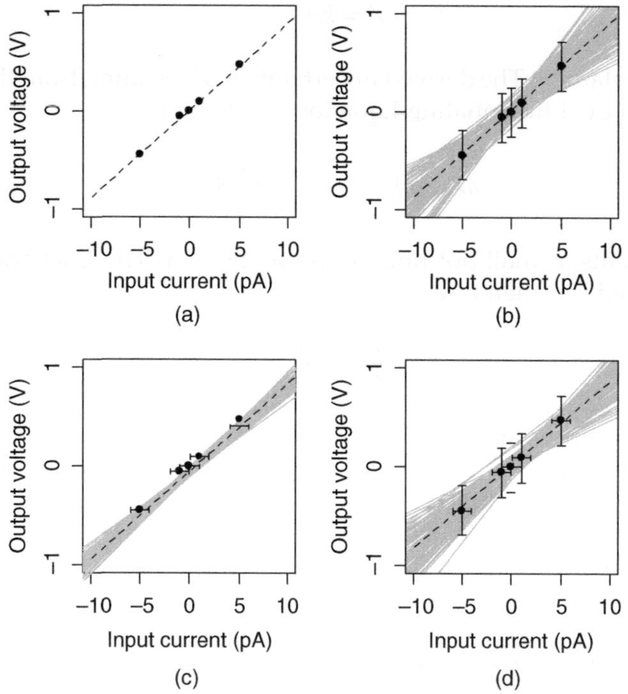

Figure 2.10 Measurements obtained from calibration of a picoammeter, through applying fixed input currents to characterise the output response. (a) Line-fit (dashed line) to the data points recorded as the best estimates of input current and output voltage. (b), (c) and (d) Range of line-fits obtained (grey lines) if the probable error range is allowed for randomly in the data points, for (b) output voltage uncertainty of $\pm$ 0.25 V, (c) input current uncertainty of $\pm$ 1 pA, and (d) with these combined uncertainties applied to input and output. (The median gradient in (d) is 84.5 mV pA^{-1} and 95% of the fitted lines have gradients between 51.2 mV pA^{-1} and 113.8 mV pA^{-1}.)

be subject to uncertainties, either because of instrument fluctuations or instrument biases, or in fact the measurements obtained have standard errors associated which are known.

Figure 2.10b considers the effect of uncertainties in the output voltage. These uncertainty ranges are conventionally drawn on the points as error bars, which show the likely range of the uncertainty, or the probable error. This indicates that there is wider range of possible fit lines which would not be inconsistent with the data, which is the case if they still pass through the vertical error bars. (Figure 2.10b shows the range of lines obtained if the error bar is regarded as representing 95% of the values possible around each point.) There may alternatively be uncertainties in the input parameters. Figure 2.10c considers the effect of uncertainties in the input current, which are shown by the horizontal error bars, and the associated range of fit lines. In general, uncertainties can be expected in both parameters, Figure 2.10d, and there are error bars in both directions, effectively surrounding each point with an ellipse, within which the more central values become more probable. By combining both sources of uncertainty, the range of possible fit lines can be estimated, and with it, the range on the instrument sensitivity obtained.

The uncertainty in a fit line can usually be obtained from statistical packages (most usually for uncertainties in one variable only, in which case, it should be the variable with the largest uncertainties), but it is also possible to estimate it subjectively by eye. In this case, the estimates of the extreme maximum and minimum line fits which pass through the error bars are chosen.[x] This of course, only provides an estimate on the gradient, but, as it is likely to be an overestimate, it is unlikely to be fundamentally misleading, although it will not use the data as effectively as an objective method, using full statistical calculations.

[x] Another subjective approach is to estimate the typical size of the residual δY (the difference between the point and the line), and to calculate $\delta Y/Y$, where Y is the span of the y variable. The fractional uncertainty in the gradient is then $\sim \delta Y/Y$.

3

Electronics and Analogue Signal Processing

Modern instrumentation is almost invariably electronic, in that measurements are obtained by a sensor providing an electrical output, either through an electrical sensor directly or via a transducer able to change the sensor's response into an analogue electrical parameter. The electrical output is most commonly a voltage, but current and frequency outputs are also used. Some digital sensors turn this electrical output into a numeric representation of the sensed value, which can be used directly by a computer. They can be considered as effectively containing a customised data acquisition system (see Section 4.2), within which the analogue signal processing stages are not accessible.

In many cases, a sensor, such as a resistance thermometer, thermocouple, ground heat flux plate, solar radiation sensor or generator anemometer, simply provides an analogue electrical variation, so it is necessary to convert this to a suitable form for a measurement system to display, record or digitise. This requires some appreciation of signal processing electronics, such as voltage amplifiers, signal converters (e.g. from current or resistance to voltage) and line drivers, which ensure the signal voltages are not degraded if they are required to pass along long connection wires to a distant measuring system. Whilst there are now complex integrated circuits which perform many of the basic signal processing requirements (or provide an array of functions allowing them to be configured by programming), the functions of these basic building blocks of signal processing still need to be understood. In some cases, such as in developing instrumentation for research applications where there is no commercial device available, or when particularly demanding performance is needed, there may be no alternative to designing new electronics at the individual component level.

Some familiarity with electronic components and their operation is necessary here, as well as the use of circuit schematic diagrams.[i] Figure 3.1 summarises the schematic symbols for some basic electronic components, which are now supplied in a wide range of packages from those suitable for *ad hoc* human construction (e.g. Figure 3.3) to those with dense connection pins intended for automatic manufacture techniques.

[i] A venerable but still excellent source of electronics wisdom is *The Art of Electronics* (P. Horowitz and W. Hill, 2nd edition, Cambridge University Press, 1989).

Meteorological Measurements and Instrumentation, First Edition. R. Giles Harrison.
© 2015 John Wiley & Sons, Ltd. Published 2015 by John Wiley & Sons, Ltd.
Companion website: www.wiley.com/go/harrison/meteorologicalinstruments

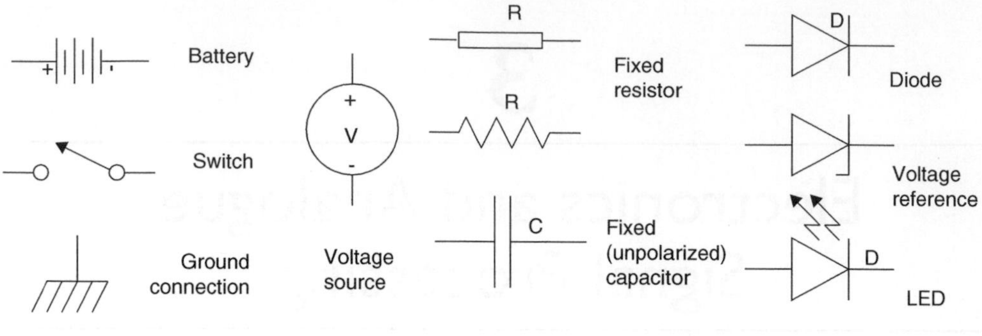

Figure 3.1 Circuit symbols for basic electronic components, of which there are very many variants. Resistors, capacitors and diodes are conventionally numbered in circuit diagrams (as R, C, D respectively), and resistors can be shown as either zig-zag symbols or rectangles. In the symbol for a diode, the arrowhead shows the direction of current flow under forward bias, from anode to cathode. Integrated circuits formed from many components are usually shown by a box which summarises their overall function and marks the relevant input, output, control and power supply connections.

3.1 Voltage measurements

A sensor providing a voltage output can be directly connected to a measuring device such as a voltmeter,[ii] as long as the range of voltages it produces lies within the range of voltages which can be measured, and the loading of the signal source itself by the measuring device is negligible. In the simplest case, the connection is straightforward, requiring just two connections (Figure 3.2). In the most general situation of an independent voltage source such as a battery or solar panel which can be considered to be 'floating' in terms of absolute voltage, the voltmeter will simply measure the voltage difference presented between its terminals. More often, however, the sensor is referenced in some way to another fixed voltage, such as the lowest voltage present. This is usually known as *signal ground*, at or close to earth potential. For a voltmeter with genuinely floating inputs, such as one which is itself battery powered, this presents no problem and the difference between the voltages will still be measured. This is a *differential* voltage measurement (see also Section 3.3.2). However, if the voltmeter does not have floating inputs, and the voltage source and voltmeter have slightly different signal ground voltages, a current will flow. This 'ground current' can affect the measurement.

Input voltages can be just a positive voltage with respect to ground (which is known as a *unipolar* input), or may swing between both positive and negative (known as a *bipolar* input).

3.2 Signal conditioning

Should a sensor not produce a varying analogue voltage, or its drive capability be insufficient for direct measurement of a voltmeter or data acquisition system, some

[ii] A voltmeter may be a handheld or bench device for temporary measurements, or in the form of a panel meter for inclusion in the enclosure of a stand-alone, single-purpose instrument. The voltmeter operation is identical, but, in the first case, the power supply of the voltmeter will usually be entirely separate from the system tested, whereas in the panel meter case, a common power supply may be shared with the sensors.

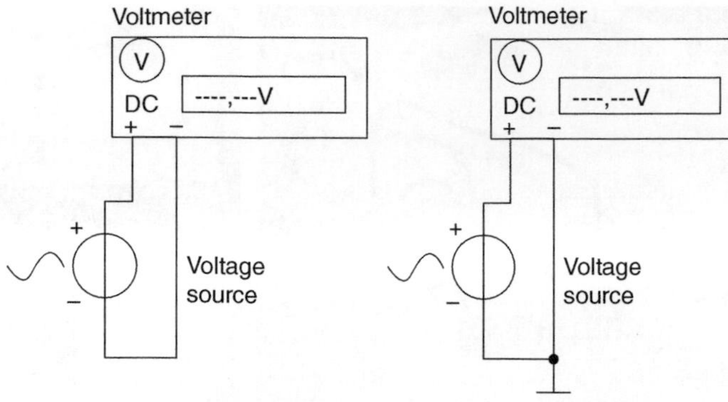

Figure 3.2 Connection of a voltage output sensor to a measuring device such as a voltmeter when the voltage source is floating (left-hand side), or grounded (right-hand side).

signal processing electronics will be required. This can take a variety of forms. The simplest signal processing is voltage amplification, through which a small signal is increased in magnitude so that it can be measured, or by which the loading of the signal source is minimised with or without amplification. In other cases, it may be necessary to convert a current to a voltage or represent a pulse repetition rate by a voltage. The exact electronics required varies, but, for the simplest case of voltage amplification, signal processing stages using standard electronic circuit building blocks can be readily implemented.

3.2.1 *Operational amplifiers*

A very widely used electronic circuit element in amplification is the *operational amplifier* (or 'opamp'). Such devices provide substantial voltage gain and are highly versatile in that they can be configured with only a few external components to function as entirely different signal processing elements. The description *operational* refers to these devices' early uses in carrying out mathematical operations for analogue[iii] computation during the late 1940s, such as integration and differentiation, and solving differential equations (see Figure 3.3). Even straightforward voltage amplification can be regarded as a mathematical operation, that of multiplication by a constant which is the amplification factor (or *gain*) of the amplifier.

Early opamps used in analogue computation were physically substantial, constructed from a pair of thermionic valves. These were superseded by opamps based on transistor devices, as semiconductors become more widely available in the 1960s. An ever-increasing range of semiconductor opamps is now available, with different devices optimised for speed, current consumption, stability, drive capability, noise and supply range, amongst many other design parameters. Their use in computational circuitry is now rare, but they can be used to linearise the responses of some sensors or compensate for the non-ideal behaviour of other components.[iv]

[iii] Analogue computation uses a replica (i.e. an analogous) electronic system to perform the calculations required; the term analogue is usually also applied to all non-digital electronics, in which signals are able to vary continuously.
[iv] An opamp can, for example, be used to remove the forward voltage of a diode, to produce an ideal rectifier.

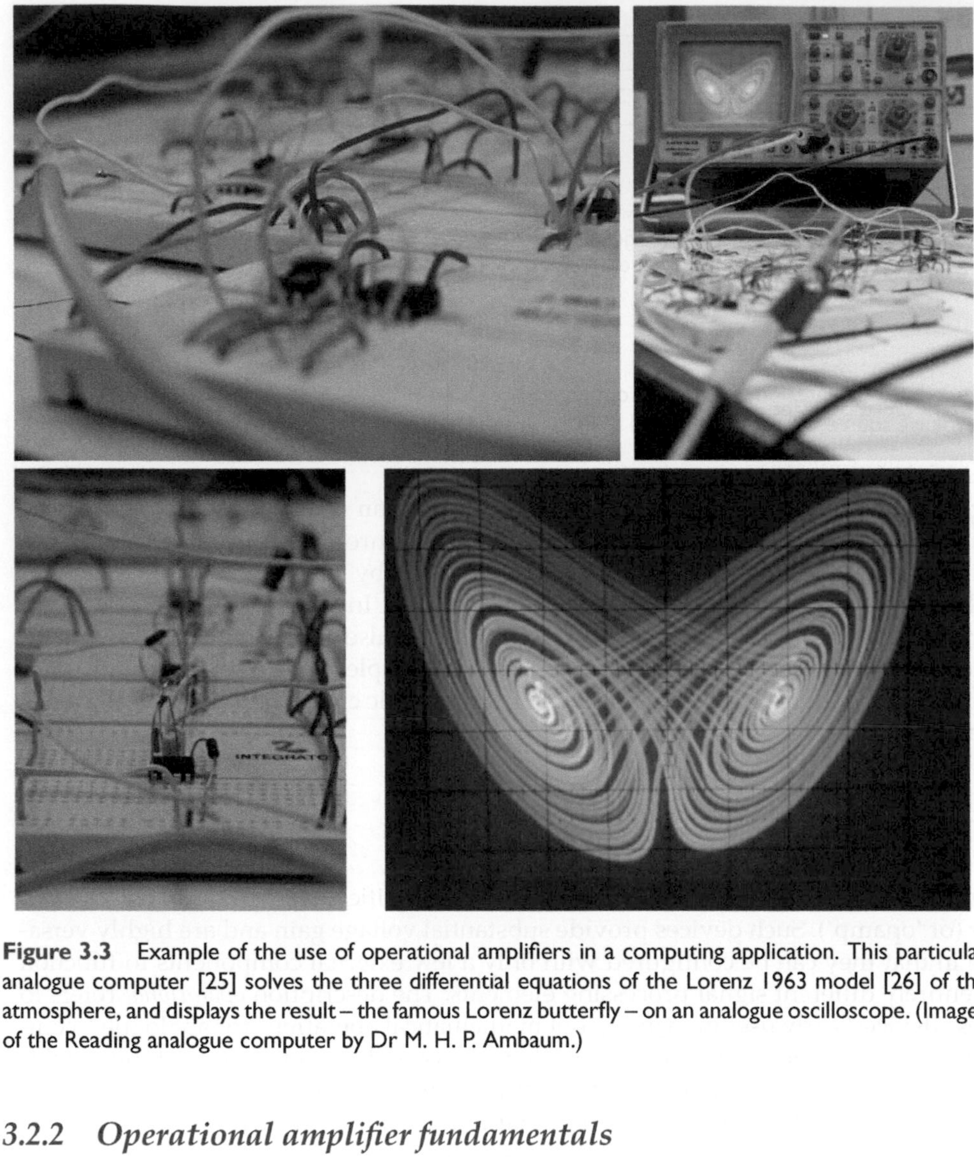

Figure 3.3 Example of the use of operational amplifiers in a computing application. This particular analogue computer [25] solves the three differential equations of the Lorenz 1963 model [26] of the atmosphere, and displays the result – the famous Lorenz butterfly – on an analogue oscilloscope. (Images of the Reading analogue computer by Dr M. H. P. Ambaum.)

3.2.2 Operational amplifier fundamentals

The typical appearance of an opamp in standard integrated circuit packaging can be seen on the plugboards in Figure 3.3 and is depicted in Figure 3.4. The central amplifying element is conventionally shown in a schematic diagram as a triangular symbol, pointing in the direction of the signal flow.

In the left-hand panel of Figure 3.4 it can be seen that the core amplification stage has three connections. Two connections (on the left of the triangle) are inputs, marked as (+) and (−) and known, respectively, as the non-inverting and inverting inputs: the third (on the right of the triangle) is the output. Internally, opamps are designed to ensure that the inputs draw very little current, to ensure minimal loading of the voltages to be amplified. Beyond the input and output connections, V_+ and V_- are the positive and negative power supply connections required to make the integrated circuit function, which will always at least encompass the signal ranges at the input

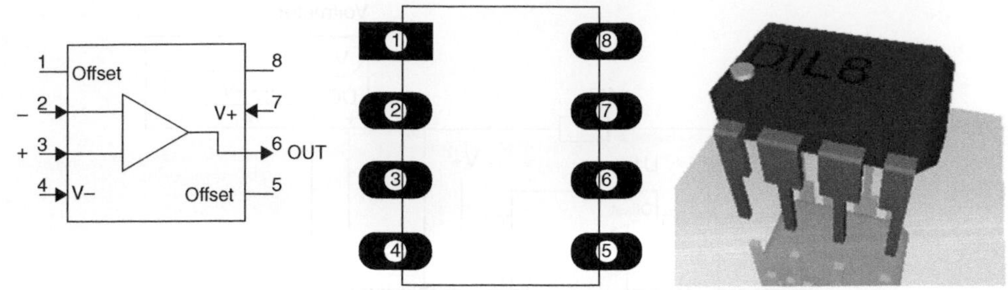

Figure 3.4 An operational amplifier, as its functional form appears in an electronic circuit diagram (left-hand panel), where the pin numbers refer to an integrated circuit chip package as viewed from above (centre). The chip package is usually marked in some way to signify pin 1 (right-hand panel). (Images from the Target 3001 electronics design package.) The arrangement of two parallel rows of connection pins spaced on a 0.1 inch (2.54 mm) matrix is known as a 'dual in line' package (abbreviated to DIL or DIP). Smaller packages suited to automatic assembly techniques are available, such as surface mount devices (SMD). Integrated Circuits are usually designated in schematic diagrams by U*n*, or IC*n*, to allow them to be identified in associated circuit descriptions.

and output. The standard power supply operating voltages are ±18 V but this large range may be unnecessary if the signal range anticipated is smaller. Other functions of the opamp's pins are specific to the particular device considered, and may allow circuitry to be connected to correct for small offset (i.e. error) voltages, or to turn the output on and off.

Without any other circuitry attached, an opamp will apply a very large amplification factor, known as the open loop gain, to the voltage difference between its two inputs. This means that if the (+) input voltage v_+ exceeds that of the (−) input v_-, the output v_o will go as positive as it can (approaching or equalling V_+), and in the reverse case ($v_- > v_+$), v_o will go as negative as possible (towards V_-). This primitive mode of operation amounts to a *voltage comparator*, as the relative magnitudes of the two input voltages v_+ and v_- yield a strongly positive or negative output, effectively a binary outcome. A special case arises if v_+ is held at 0 V, as then a v_- voltage, more positive or more negative than 0 V, will lead to the opposite polarity voltage at v_o. This operation provides a sign inversion.

Rather more effective versions of an *inverter* circuit are present in digital integrated circuits, which, when presented with input signals representing the 0 or 1 binary condition according to some pre-arranged voltage thresholds, generate the opposite condition at their output. These are commonly used for frequency generation (see also Section 3.6.1), or as part of converting voltages to a digital representation (Section 4.1.4).

3.2.3 *Signal amplification*

An opamp is, however, almost never used with its full open loop gain, and additional components are used to restrict the gain to a defined value. This ensures a linearly varying output rather than a solely digital response.[v] Figure 3.5 shows the additional

[v] *Comparators* are also available as specialised devices, optimised for fast switching and output drive capability.

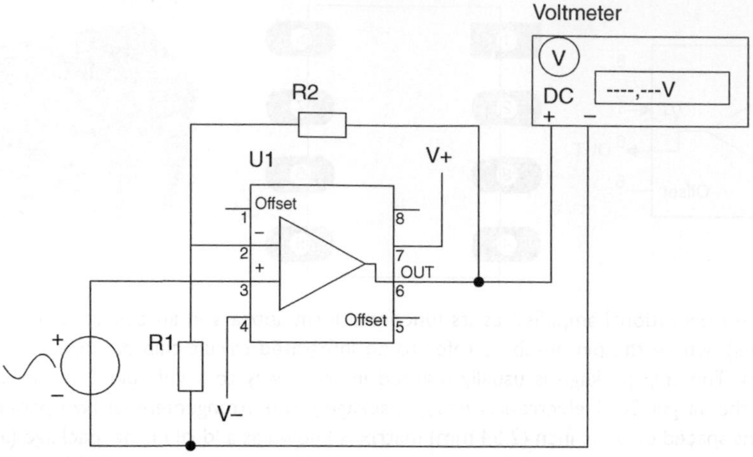

Figure 3.5 An operational amplifier (U1) wired as a non-inverting amplifier. The voltage presented at the non-inverting input is amplified by an amount determined by the resistors R1 and R2. (Offset terminals are available on some operational amplifiers, allowing the device's offset voltage to be corrected using external components connected to these pins.)

circuitry required to configure the opamp as a non-inverting amplifier. One resistor (R1) is connected from the inverting input to signal ground, and a further resistor (R2) is connected between the output and the inverting input. R2 is the *feedback* resistor, as a fraction of the output voltage is applied (fed back) to the input. In this configuration, the voltage gain G is defined by

$$G = 1 + \frac{R_2}{R_1}. \tag{3.1}$$

That is, the output voltage v_o at the voltmeter and the source voltage v_i will be related by $v_o = Gv_i$.

For correct operation, it is essential that the supply voltages to the opamp (V_+ and V_-) are greater than the input or output signals expected, and within the maximum range allowed for the actual device in use. This operating range is known as the common mode range. Opamps vary in their behaviour when the input (or output) signals are close to the power supply rail voltages V_+ and V_-. Some devices are said to be 'rail to rail' if they are able to accept or produce voltages all the way to the supply voltages; other devices only show linear operation close to V_- and still others can only operate properly with input and output voltages which are well within the supply rails. So-called 'single supply' opamps are optimised to allow linear operation down to the negative supply. This means that, if the signal is positive-going only, the signal ground can be used as a power supply connection and no negative power supply is needed. Availability of power supplies therefore presents an important consideration in choosing an opamp for a particular application: a dual (positive and negative) supply is more costly and complicated to implement than a single (positive only) supply.

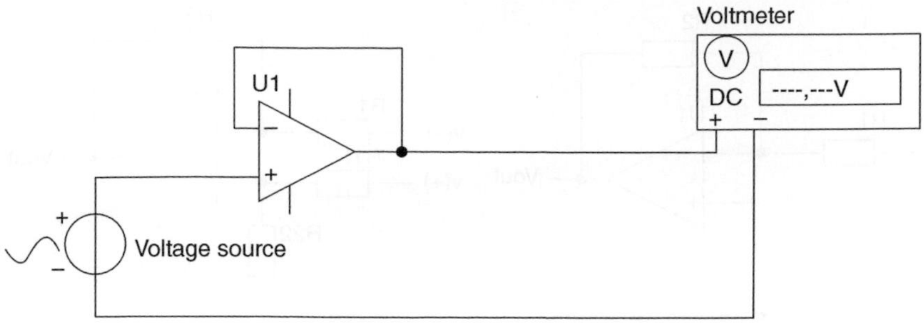

Figure 3.6 Unit gain ('buffer') amplifier, in which the voltage source is not amplified by U1 but, because the circuit configuration provides minimal loading, allows sensitive signal sources to be measured with a conventional voltmeter. (Power supply connections are not shown.)

3.2.4 *Buffer amplifiers*

An important special case of the non-inverting amplifier arises when the output and input are connected (i.e. R2 = 0) but no connection is made to signal ground (R1 → ∞), as shown in Figure 3.6. From Equation 3.1, such a circuit will have a gain $G = 1$. Although this clearly provides no voltage gain, such a 'unity gain buffer' (or voltage follower) is a highly useful first stage for many sensitive signal sources.[vi] This is because, as the opamp input usually only draws negligible current, loading of the signal source by the measuring system is minimised. Other than offset (error) voltages or other departures from ideal opamp behaviour, for example, at high frequency, the output voltage presented to the voltmeter will closely replicate the signal voltage.

Important considerations in implementing buffer amplifiers are to ensure that the opamp is selected for low input current (also known as *bias current*) and that the signal range does not extend beyond the supply voltages. Opamps are constructed internally from transistors of different kinds, but some of the lowest bias current opamps employ internal Field Effect Transistors (FETs), which have particularly restricted supply voltage ranges. This reduces their usefulness as buffer amplifiers in the basic form of Figure 3.6, but, as their low input bias current may be an overriding requirement in allowing a measurement, techniques [27] exist which nevertheless allow restricted supply opamps to measure wide input voltage ranges[vii] (see also Section 3.3.1).

3.2.5 *Inverting amplifier*

As well as voltage amplification in which the sign of the input voltage is preserved at the output, an input signal can be applied to the inverting input. If this input is made through an input resistor, and feedback applied from the output to the inverting

[vi] Limited current (or 'high impedance') signal sources include ion probes, electrochemical sensors (for gas measurements), pH probes or high resistance thermistors.

[vii] These approaches work by ensuring that the opamp itself is never exposed to more than a small voltage difference, by using it to compare the input voltage with a locally generated voltage which it can adjust.

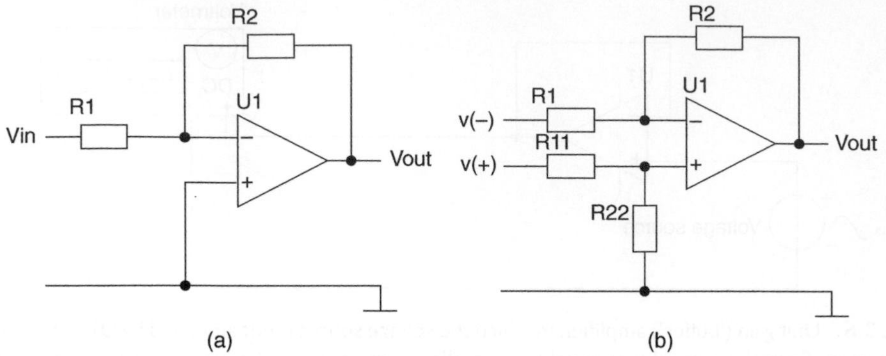

Figure 3.7 Circuitry associated with an opamp (U1) for (a) an inverting amplifier and (b) a differential amplifier. (Power supplies are not shown in either case.)

input, the circuit of Figure 3.5a – an *inverting amplifier* – results. The voltage gain of this circuit is

$$G = -\frac{R_2}{R_1} . \tag{3.2}$$

If high gain is needed, the ratio required of R2:R1 may be such that R1 becomes sufficiently small to load the sensor connected (or the previous stage), and it may be necessary to use two stages. Another application of Figure 3.7a is for resistance measurement (see also Section 3.5), as, if V_{in} is fixed and R1 a known precision resistance, V_{out} will vary linearly with R2.[viii]

An extension of the inverting amplifier circuit is to present input signals to both opamp inputs, which forms a *differential amplifier* (Figure 3.7b. This amplifies the difference in input voltages giving an output voltage v_o according to

$$v_o = (v_+ - v_-)\frac{R_2}{R_1} , \tag{3.3}$$

where R22 = R2 and R11 = R1. The performance of the differential amplifier circuit is critically sensitive to imbalance in the matching of the resistors, and the other resistances associated with the voltage sources to be measured. One property also affected by resistor imbalance is the *common mode rejection ratio* (CMRR), which defines the extent to which an amplifier's performance varies with the size of its input signal within its common mode.

To optimise performance, each input voltage can be first applied to a unit gain buffer before the differential stage. Such a hybrid configuration of several opamps is known as an *instrumentation amplifier*, as it can operate as inverting, non-inverting or differential with the gain required set by external resistors. Some instrumentation amplifiers also allow the gain to be selected from a range of accurate pre-installed values, as closely-matched resistors can be fabricated on chip.

[viii] By careful choice of values, for example V_{ref} = 1.0 V and R1 = 1 kΩ, the linear relationship between V_{out} and R2 can have a straightforward scaling of (−)1 V per kΩ, forming a simple Ohmmeter.

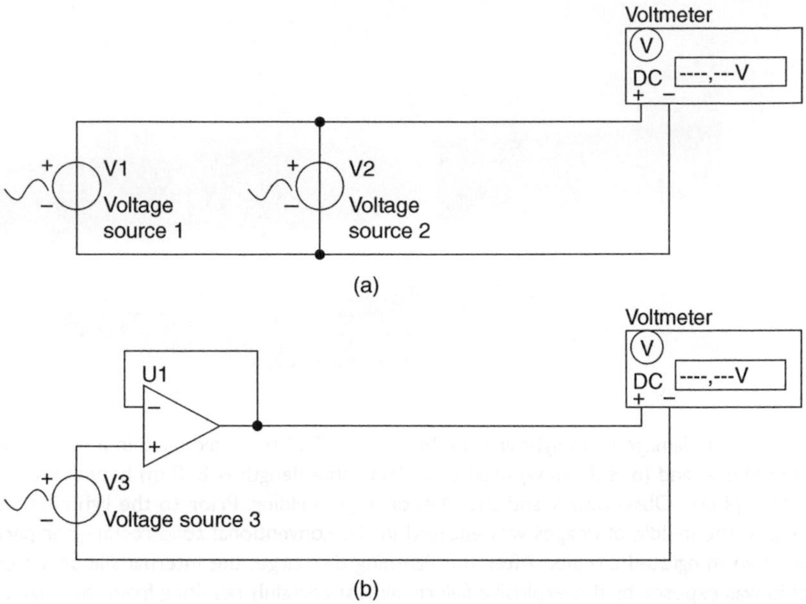

Figure 3.8 Effect of long connection wires on voltage measurements. (a) Weak signals from voltage source (V1) may be swamped by error voltages from another voltage source, (V2). (b) An improvement is obtained if a weak voltage source (V3) can be buffered close to the source before connecting to a long signal line.

3.2.6 *Line driving*

In real environmental measurement situations, it may be unavoidable that a sensor has to be situated at some distance from the measurement system. This can lead to signal degradation and poor quality measurements, particularly if the signal source has very limited or no current drive capability. In such circumstances screened cable[ix] is often helpful but, for some signal sources with almost no current driving capability, the cable itself can still present an unacceptable load on the sensor. Figure 3.8 shows a situation which can arise with long cables. In Figure 3.8a, a weak voltage source connected by long cables to a voltmeter can be influenced by error voltages arising, perhaps induced by an electrically noisy environment, induced radio signals or as a result of thermoelectric effects at dissimilar metal junctions (see also Section 5.3.1). In Figure 3.8b, the signal is buffered close to its source, and the line is instead driven actively by the opamp. This is preferred as it reduces pick-up of error voltages on the line.

In order for Figure 3.8b to work correctly, however, it is also essential that the output drive capability of the buffer stage is sufficient for the connections required. Long connection cables present a complex load formed from the line capacitance and inductance which can introduce phase distortion. Special opamps may therefore be needed to drive appreciable load capacitance. These are internally stabilised (or compensated) to remove the risk of oscillation, which may otherwise arise at the frequency

---

[ix] Screened cable consists of a central conductor, insulated and surrounded by an outer conductor of wire mesh or tape. The outer conductor acts as a screen which is usually connected to signal ground.

(a) (b)

Figure 3.9 Physical damage from lightning on 28 January 2004 to a chip used in a digital line driver stage ((a) from above, and (b) side view) used on a data cable (length ~ 800 m) between the Reading University Atmospheric Observatory and the Meteorology Building. Prior to the lightning discharge, the driver chip in the middle of images was encased in the conventional solid rectangular package, as for the other two integrated circuits. After the lightning discharge, the internal silicon wafer of the line driver chip was exposed by the explosive failure, almost certainly resulting from the huge transient power dissipation.

when the phase lag leads to exact constructive interference between input and output (see also Section 3.6.1).

A further consideration with analogue or digital line driving stages is in avoiding the damage which can occur from induced transients on the cables connected, for example, from lightning. For this reason, long cable lengths are again undesirable, and the use of energy absorbing components combined with voltage transient limiters can help to prevent the catastrophic damage shown in Figure 3.9.[x]

3.2.7 *Power supplies*

Amplifier circuits require unipolar or bipolar power supplies, which add further complexity to the signal conditioning system. In the simplest case when a good quality regulated low voltage power supply is available, all that may be needed is a direct power connection to the signal conditioning circuit. Even then, however, it is important to ensure that noise in the power supply does not influence the analogue signal circuitry. Low voltage power supplies from the AC mains supply are usually derived through transformers or by high frequency step-up or step-down 'switching' systems, both of which tend to generate noise, which in turn may find its way into the signal path. Other devices connected to the same low voltage supply, particularly computers, can also generate interference through the power supply. In general, therefore, further power supply conditioning in an instrument is usually desirable, both to regulate and filter the power supply connection. This serves to stabilise the supply voltage regardless of wide variations in the currents drawn, and to reduce high frequency noise present on the power connections. Filtering is usually

[x] An alternative approach which avoids long signal connections is to convert the signal to digital form and send the digital representation optically or wirelessly.

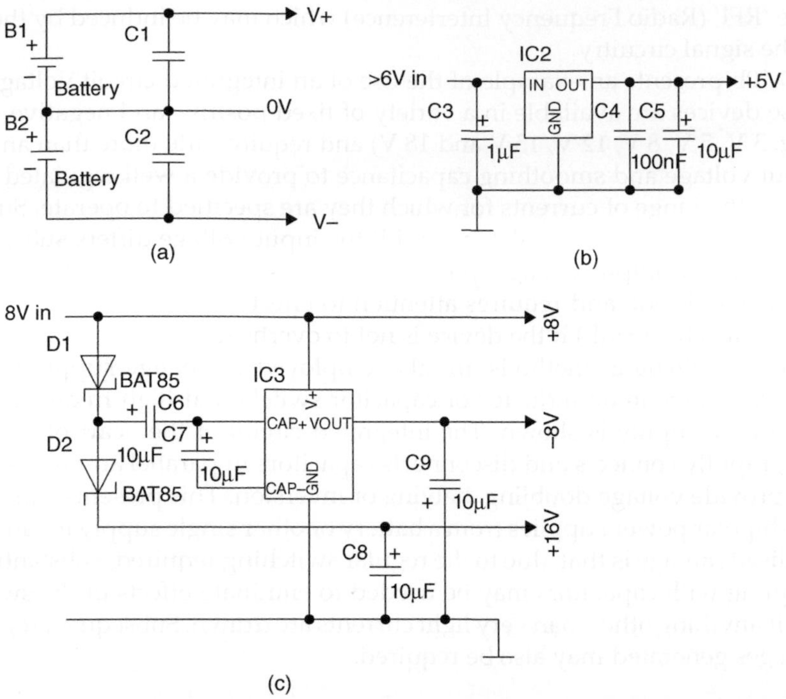

Figure 3.10 Power supply circuits. (a) shows a bipolar power supply for a handheld instrument using two batteries wired in series, to provide V_+, V_- and 0 V. (b) shows a 'three-terminal' (input, output and ground) linear voltage regulator circuit, to provide a stabilised power supply. Such regulators are available for a wide range of positive or negative voltages (typically 3 V, 5 V, 9 V, 12 V, 15 V, 18 V and 24 V). For a +5 V supply as shown, a type 2950 device for IC2 provides little voltage drop and current consumption. (c) shows a 'flying capacitor' switching supply (using IC3, a type 7660 chip), which, from a 8 V supply, generates −8 V and 16 V by parallel and series switching of the capacitors C6 and C7. Further regulation or filtering may then be required, such as that provided by C8 and C9.

implemented using capacitors, which, wired in parallel across power supplies, provide a low impedance path for high frequencies. The use of a series inductance offers an alternative or complementary approach, as this inhibits high frequencies from passing down a connection line whilst still allowing DC currents to flow.

Figure 3.10 shows power supply circuits suitable for powering signal conditioning circuitry. In Figure 3.10a, two batteries are used as they might be in a handheld instrument. Decoupling capacitors in parallel with the supply lines act to reduce high frequency noise on the supply which may otherwise be propagated through to signal voltages. The capacitance required depends on the frequencies present. If digital circuitry is also used, such as in the form of a microcontroller (see Section 4.2), it is usually necessary to include decoupling close to its supply connections as well. For a circuit in close proximity to a radio transmitter, an LC filter[xi] can be needed to[xii]

[xi] The combination of a series inductor L and a parallel capacitor C forms a low pass filter with cut-off frequency f_c where $f_c = 1/[2\pi\sqrt{(LC)}]$ which has little effect on low frequencies ($f \ll f_c$), but attenuates high frequencies ($f \gg f_c$).

[xii] At high frequencies very little filtering inductance is needed to remove RFI, and merely adding inductance by sliding ferrite beads along existing cables or passing the cables through ferrite rings can be effective.

combat the 'RFI' (Radio Frequency Interference) which may be induced by the transmitter in the signal circuitry.

Figure 3.10b presents an example of the use of an integrated circuit voltage regulator. These devices are available in a variety of fixed positive and negative voltage values (e.g. 3 V, 5 V, 8 V, 12 V, 15 V and 18 V) and require little more than an unregulated input voltage and smoothing capacitance to provide a well regulated output voltage, over the range of currents for which they are specified to operate. Such regulators are known as linear regulators, and if the input voltage differs substantially from the regulated output voltage, power will be dissipated as heat in the regulator, which is inefficient, and requires attention to effective thermal transfer (e.g. by radiation, using a heat-sink) if the device is not to overheat.

Alternative 'switching' methods are also employed to change supply voltages, using energy storage in an inductor or capacitor switched at high frequency. In (c) a switching-type supply is shown. The integrated circuit at the heart of the circuit (type 7660) rapidly connects and disconnects capacitors in parallel and series combinations to provide voltage doubling, halving or inversion. This presents a method of producing bipolar power supplies from a battery or other single supply for an opamp circuit. A disadvantage is that, due to the regular switching required, substantial supply decoupling with capacitors may be needed to eliminate effects of the switching frequency if anything other than very light currents are drawn. Subsequent regulation of the voltages generated may also be required.

3.3 Voltage signals

Voltage processing is commonly required across many meteorological applications for the signal conditioning associated with thermocouples and thermopiles, such as in radiation measurement and for some configurations of resistance thermometer. An even more basic atmospheric voltage measurement is in determining the atmosphere's electric potential at a known height, as in fair weather, the potential at 1 m above the surface is about 100 V.

3.3.1 Electrometers

Measuring the atmosphere's electric potential almost certainly presents the most fundamental meteorological application of a voltage amplifier, as the quantity required is itself a voltage. In principle, nothing more than a buffer amplifier is required, combined with a suitable antenna to acquire the atmosphere's potential at a particular height. A typical antenna could be a long horizontal wire, supported at each end by high quality insulators mounted on vertical posts (Figure 3.11). Under fair weather conditions with good turbulent mixing, such an antenna will charge positively [28].

Such an antenna arrangement can supply hardly any current (< 0.1 pA), and consequently an ultra-low current (or *electrometer*) buffer amplifier is required. Powered from a 9 V battery, with the negative connection to earth, an electrometer buffer would allow antenna potentials between 0 and 9 V to be measured. Unfortunately, even for a sensing antenna suspended at only 10 cm above the surface, this will be insufficient in dynamic range, as the antenna potential will then usually be at least 10 V.

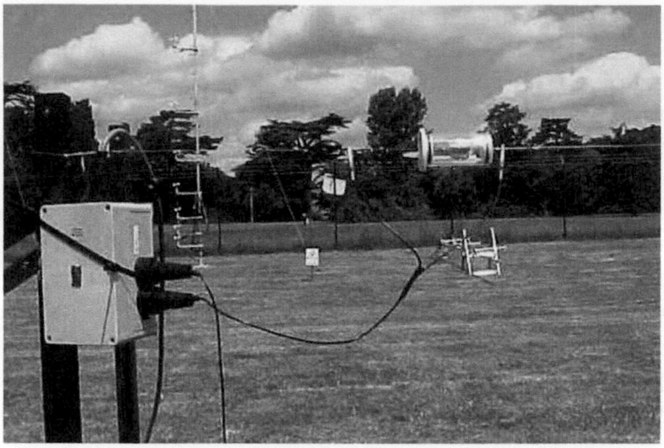

Figure 3.11 Horizontal wire antenna (right-hand side of photo) installed at Reading University Atmospheric Observatory, for measurement of the atmospheric potential. (The electrometer for measuring the potential is mounted in the weatherproof box on the left.)

The operating range of the simple electrometer buffer can be extended with an auxiliary amplifier [29], as shown in Figure 3.12. A buffered version of the antenna potential is provided as the output voltage. In this circuit, U1 is the critical electrometer buffer amplifier, again powered by a 9 V battery but with a supply splitter added to generate the battery midpoint voltage G0. If the battery-powered U1 buffer circuit alone were connected to the antenna, but with no connection made to earth, the whole circuit would eventually acquire (or 'float' to) the antenna's potential. The potential with respect to earth however is required, for which a further buffer amplifier (U2) is used. This amplifier has earth potential (0 V) within its power supply range. Because the weak source of antenna potential has already been buffered by U1, U2 does not need the electrometer performance and hence the selection constraints on U2 are less critical. An opamp type allowing the standard ±18 V supply rails (or greater) can be

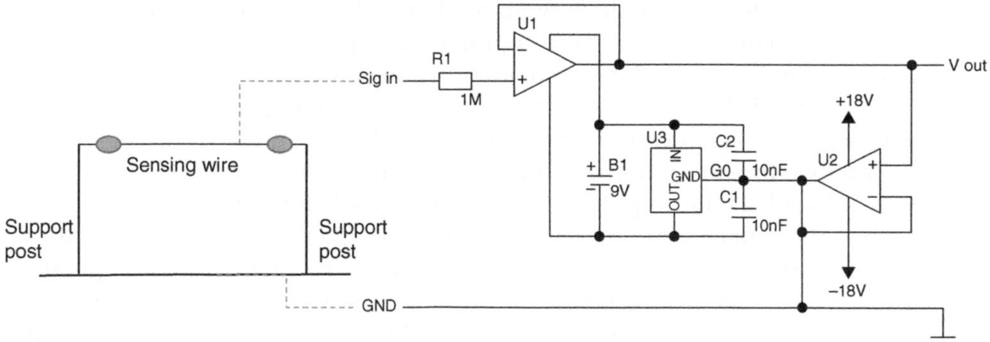

Figure 3.12 Schematic of an extended range electrometer buffer amplifier used to measure the atmospheric potential, as acquired by a long horizontal wire strung between insulators and support posts, with respect to earth potential. U1 is the buffer (electrometer) amplifier, U2 an auxiliary amplifier and U3 a power supply splitter, which generates a voltage G0 at the midpoint of the 9 V power supply from B1.

chosen. As U2 is also configured as a unit gain buffer, its output will have the same potential as the output of U1, which is that of the antenna. Connecting U2's output to point *G0* still allows U1's potential to float, but ensures this becomes referenced to earth. This arrangement allows U2 to adjust the midpoint of U1's power supplies, as the antenna potential varies, allowing the antenna potential to remain within the operating range of U1.

A useful feature of the auxiliary amplifier/electrometer buffer combination is that, whilst it ensures the rather restrictive requirements of the electrometer stage are met, other critical performance parameters such as voltage range and CMRR, which can both be quite poor for electrometer operational amplifiers, are transferred to the auxiliary amplifier. By fixing the electrometer stage's input voltage with respect to its local power supply with the auxiliary amplifier, the electrometer stage's performance becomes more consistent over the whole input voltage range of the combined circuit.

3.3.2 *Microvolt amplifier*

Some environmental sensors, such as thermocouples or thermopiles, generate low level outputs at the tens of microvolt level or smaller. Accurate amplification of these small voltages is a critical aspect in obtaining good quality measurements from the sensors concerned, and attention is given to avoiding generating further voltages which will contribute to the overall uncertainty. The voltage developed across a thermopile is a floating signal source, and hence a differential measurement is appropriate. Differential amplifiers can be constructed from opamps as indicated in Section 1.2.5, but, because general purpose opamps typically have offset voltage errors ~1 mV, these can easily swamp the signal voltage sought if it is small. Instrumentation amplifiers exist which are optimised for microvolt level voltage applications, which may also include well-defined gain stages.

Figure 3.13 shows a differential input microvolt amplifier, intended for thermopile measurements under atmospheric conditions [30]. IC1 is a fixed gain (×100) differential amplifier, chosen for good thermal stability.[xiii] Both differential inputs are tied to signal ground through a high value resistor to minimise loading on the signal source, although, if the resistors are made very large, they will contribute additional noise voltages. The output of IC1 is passed through an RC low pass filter,[xiv] to reduce effects of high frequency noise, and IC2 then provides further gain as required, from its input of 10 mV or greater. IC2 is also chosen to be capable of driving the load capacitance associated with long signal cables. This allows the output voltage to be measured remotely, and therefore the amplifier itself can be mounted close to the signal source. To allow use of a substantial supply voltage to compensate for voltage dropped with distance along the connection cables, IC3 and IC4 provide power supply regulation at the signal amplifier. IC3 is a three-terminal linear regulator which regulates the supply to +6 V, and IC4 generates a −6 V supply using the flying capacitor technique. Smoothing capacitors C7, C8 and C9 reduce the noise on the supply, and, close to the sensitive integrated circuit IC1, supply decoupling capacitors (C1 and

---

xiii This particular differential amplifier (type LT1100) has an offset voltage drift of 5 nV K^{-1}, and a gain drift of ±4 ppm K^{-1}.
xiv A series resistance *R* and a parallel capacitor *C* yields a low pass filter with cut-off frequency f_c where $f_c = 1/[2\pi RC]$ which is less costly in space and components than a LC filter but less effective. As for a LC filter, it has little effect on low frequencies ($f \ll f_c$), but restricts the passage of high frequencies ($f \gg f_c$). (Reversing *R* and *C* yields a high pass filter).

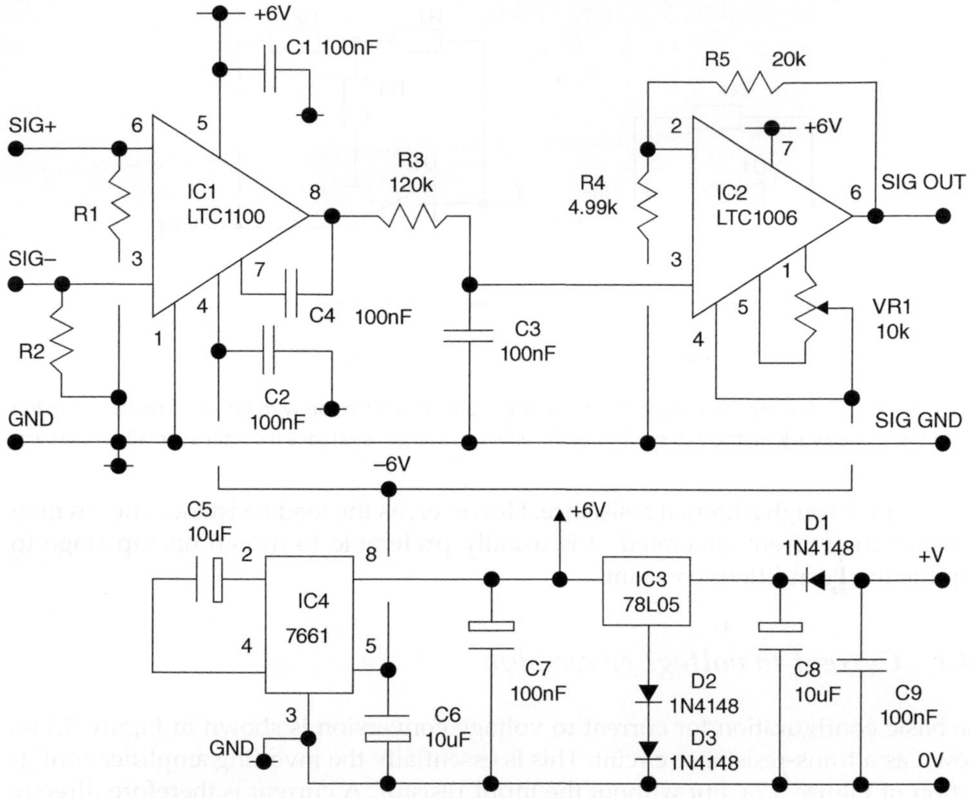

Figure 3.13 Microvolt amplifier providing ×500 gain (at SIG OUT and SIG GND) from the differential voltage at its input (SIG+ and SIG−). IC1 is the differential amplifier stage with a fixed gain of 100. IC2 is non-inverting amplifier with a gain of 5. IC3 and IC4 provide bipolar power supplies of ±6 V. (Reproduced with the permission of the American Institute of Physics.)

C2) are included. A final detail is an input diode, D1. This protects the circuit against inadvertent supply reversal, which can cause total destruction of semiconductors. The combination of protection diode and linear regulator provides good protection of the sensitive signal processing stages against supply reversal and over-voltage.

In use, such an amplifier would be mounted close to the sensor in a screened box (i.e. a metal box which itself was connected to signal ground), with short input connections from the sensor using two single core or one dual core screened cable. It can also be useful for the two insulated input connections to be twisted together. This helps to ensure that any induced voltages are common to both inputs, and they will therefore be substantially removed by the differential measurement, which subtracts one input from the other.

3.4 Current measurement

Some sensors produce currents directly, such as a photodiode used for measuring radiation or an electrochemical sensor measuring ozone on a radiosonde. The simplest method of measuring a current is to first convert it to a voltage, by passing

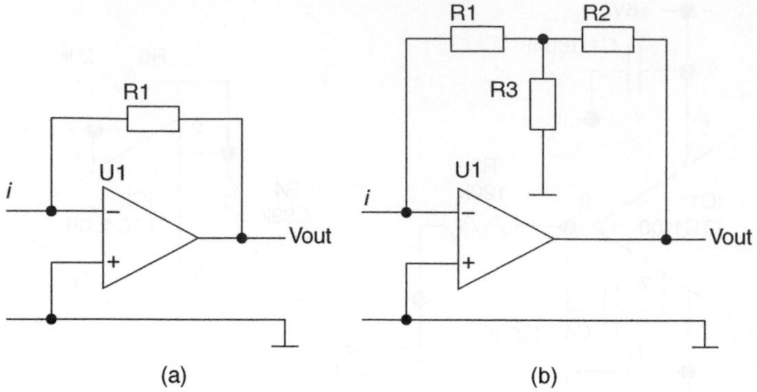

Figure 3.14 Trans-resistance stages for current to voltage conversion with (a) single resistor version and (b) the 'T' network method to synthesise the effect of a large resistance from three smaller resistors.

the current through a known resistance. However, as the load resistance chosen may influence the current generated, it is usually preferable to use an opamp stage to keep the load conditions constant.

3.4.1 *Current to voltage conversion*

The basic configuration for current to voltage conversion is shown in Figure 3.14a, known as a trans-resistance circuit. This is essentially the inverting amplifier configuration of Figure 3.7a, but without the input resistor. A current is therefore directly applied to the inverting input. The output voltage is related to the input current by

$$v_o = -iR_1 \,. \tag{3.4}$$

For good accuracy, it is important to ensure that the bias current of the opamp is much smaller than the measurement current, and it is often worthwhile to add parallel capacitance to R1 to reduce the gain at high frequencies if the current source varies. The signal source is effectively driven to signal ground by the use of the opamp, and hence R1 can be varied without affecting the properties of the source. For currents of nA and larger, R1 ∼ MΩ will give voltages in the mV range which can be amplified further if needed. For smaller currents, in the pA and fA range, R1 may need to be very large.[xv]

Resistors with values above 100 MΩ are relatively difficult to obtain, and values from 1 GΩ to 1 TΩ (10^9 Ω to 10^{12} Ω) are specialist and delicate components. An alternative configuration of Figure 3.14b is therefore sometimes used. This uses a 'T' shaped resistance network with the opamp feedback to synthesise a larger feedback resistance, giving a response of

$$v_o \approx -iR_1\left(1 + \frac{R_2}{R_3}\right) = -ir_T R_1 \,. \tag{3.5}$$

The extra resistors increase the effective feedback resistance to many times greater than that of R1 alone, but opamp offset voltage errors are also amplified by the same

xv $1\mu A = 10^{-6}$ A; $1nA = 10^{-9}$ A; $1pA = 10^{-12}$A; $1fA = 10^{-15}$ A;

factor r_T. This will exacerbate the effects of thermal drift, so as large a value of R1 as can be obtained is still preferable. For time-varying currents, the T-network approach allows a flatter frequency response by compensating for the parallel capacitance (C1) of R1, if R3 is replaced by a capacitor (C3), chosen such that the time constants are matched, that is R1·C1 = R2·C3.

3.4.2 *Photocurrent amplifier*

An example of current measurement using an opamp is in the use of a photodiode (see also Section 9.3.3) to measure solar radiation, with the signal-conditioning electronics employed shown schematically in Figure 3.15.

In the circuit [31] of Figure 3.15, the photodiode (D1) is effectively reverse-biased (i.e. its cathode voltage at the opamp's inverting input is more positive than that of its anode, at 0 V), and allows a (photo)current i_{PD} to flow out of the opamp to 0 V. This is converted to a voltage by the opamp and feedback resistor R2, and the output voltage is halved by the potential divider R3-R4, to ensure it falls within a 0 to 5 V range suitable for a subsequent measurement circuit. A subtlety is the use of a voltage reference, D2, to provide a fixed voltage (V_{ref}) above ground at the non-inverting input, in this case, with V_{ref} = 1.2 V. This serves to raise both opamp inputs to 1.2 V which both reverse biases D1 for correct photodiode operation and ensures that the opamp output with zero photocurrent remains positive. Effectively, this raises the opamp input to 1.2 V, making, in relative terms, 0 V appear as a negative supply rail, but without the complexity of a dual rail power supply. The response of the circuit to the photocurrent i_{PD} is given by

$$V_{out} = \frac{R_4}{R_3 + R_4}(i_{PD}R_2 + V_{ref}) , \quad (3.6)$$

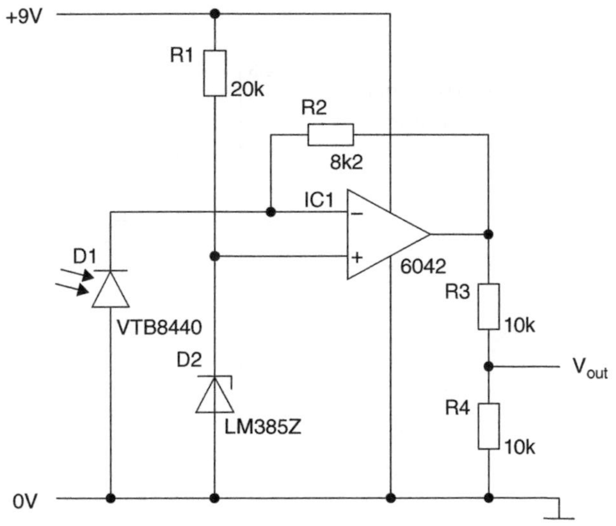

Figure 3.15 Single-supply circuit to amplify the photodiode current, to measure intercepted solar radiation. D1 is the detector photodiode, and D2 is a voltage reference, which provides 1.2 V offset at the non-inverting input. (Reproduced with the permission of the American Institute of Physics.)

which, for a linear response in photocurrent to solar radiation, indicates a linear response in output voltage.

3.4.3 *Logarithmic measurements*

If the current to be measured extends over a wide dynamic range, for example spanning several orders of magnitude, there is a risk that the signal conditioning circuit can saturate at its upper or lower limit and no measurement will be obtained. This problem can arise in the measurement of photocurrents in the atmosphere which span a large range of values during daytime, and in atmospheric electricity measurements where currents generated in elevated points vary between μA in disturbed weather to pA in fair weather. An alternative approach for current measurement is therefore needed if, in preserving the capability to measure large currents, good resolution is still required at the low current end of the range.

A wide operating range can be achieved with a logarithmic current amplifier, which generates an output voltage proportional to the logarithm of the input current. This can be implemented with an opamp by using a device having an exponential voltage-current response in place of the feedback resistor of the trans-resistance configuration, such as a diode. Light emitting diodes (LEDs) can be used as feedback elements providing such a response characteristic [32], over many orders of magnitude, as long as they are kept in darkness to prevent photocurrent generation. (There is, however, also a temperature response in these devices, but this can be compensated with further signal processing electronics using a thermistor temperature measurement [33].) The wide range logarithmic response of a current amplifier using LEDs as the feedback element is shown in Figure 3.16, for an instrument designed to measure the atmospheric point discharge current [34], which, as it is exposed to atmospheric temperature variations, requires thermal compensation.

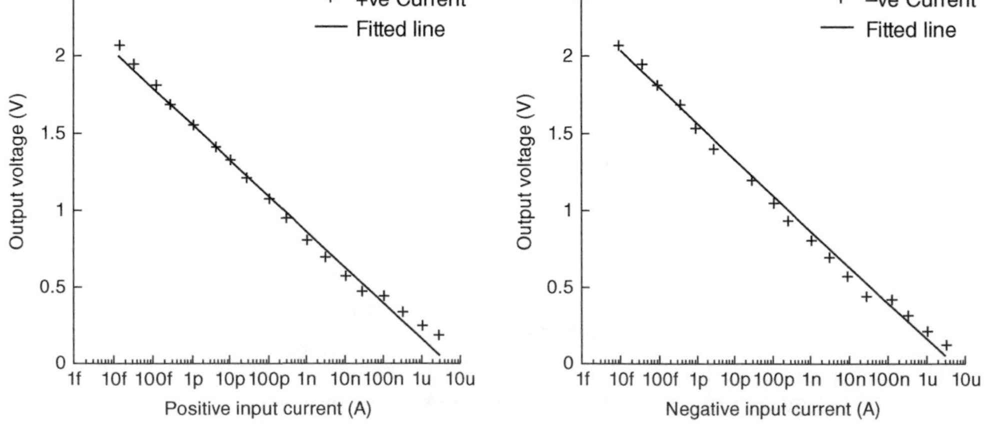

Figure 3.16 Laboratory response of a logarithmic current measuring device for the atmospheric Point Discharge Current (PDC), for applied negative and positive currents over a wide range.

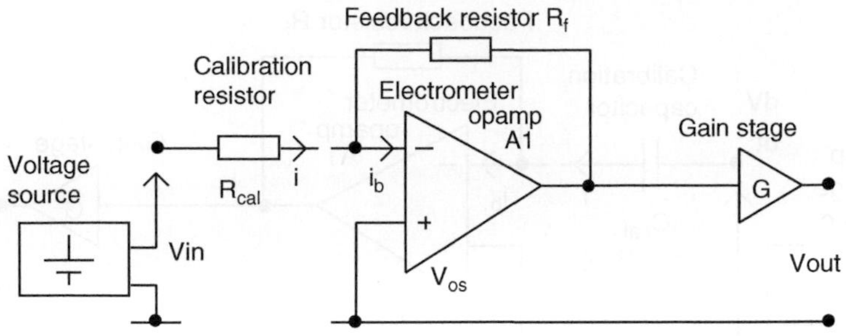

Figure 3.17 Resistive generation of a calibration current from a voltage source presenting an input voltage V_{in}. (The opamp offset voltage is V_{os}.)

3.4.4 *Calibration currents*

Current amplifiers can be calibrated by using a voltage source, applied to a known[xvi] value series resistor R_{cal} (see Figure 3.17). As the feedback current amplifier input is effectively at earth potential (known as 'virtual earth'), the whole potential difference of the voltage source is applied across the resistor.

Assuming an ideal system with a finite input-bias current i_b, the output voltage for an input current i will be

$$V_{out} = G\{[-(i + i_b)R_f + V_{os}]\}\ , \tag{3.7}$$

where V_{os} is the offset voltage of amplifier A1, typically a few millivolts.[xvii]

An alternative method of generating a small calibration current is to apply a steadily rising or falling voltage to a calibration capacitor (Figure 3.18).

This method generates the reference current by differentiation, using a precise, slowly varying linear ramp voltage, presented to a low value high quality capacitor. The current generated i is proportional to the rate of change of voltage $V(t)$, i.e.

$$i = C_{cal}\frac{dV}{dt}\ . \tag{3.8}$$

There is, however, actually a resistive component to any practical capacitor, so that the current generated will vary with time according to

$$i(t) = C_{cal}\frac{dV}{dt} + \frac{V(t)}{R_{cal}}\ , \tag{3.9}$$

if the resistive current generation is comparable with that from the capacitor. The resistive contribution can be minimised by using a small ramp voltage, such as a few

---

[xvi] This large resistance can either be measured, or its value inferred by fabricating the resistor from a series connection of known resistors. In fact, a series connection of nominal, but unmeasured, resistors has some benefit through averaging, as, for N resistors of one value in series, the total series resistance is equivalent to their mean resistance multiplied by N.

[xvii] This is the magnitude of the voltage required at A1's *input* to make its *output* voltage zero when there is no input.

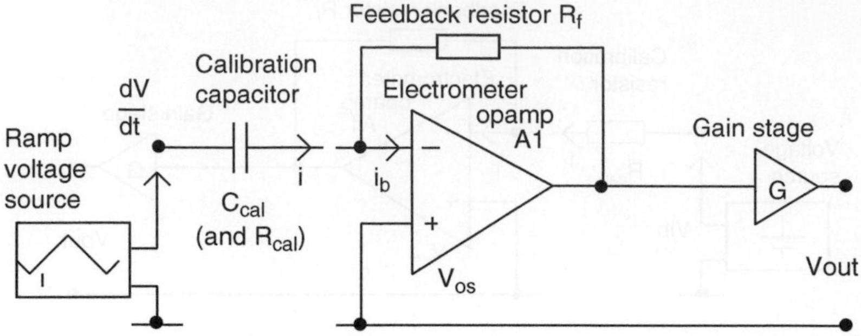

Figure 3.18 Generating a calibration current using a steadily rising or falling current applied to a calibration capacitor C_{cal}.

tens of mV per second with a good quality polystyrene capacitor. Using this approach, reference currents can be generated in the range of 100 to 1000 fA, which, if the voltage ramp is symmetrical, can provide stable and symmetrical bipolar currents [35], such as for use with balloon-carried current measurements [36].

3.5 Resistance measurement

Electrical resistance is a further essential quantity required accurately for atmospheric applications, such as in determining the resistance of a thermistor used for air temperature measurement. As electrical connections have to be made to such a sensor to measure the resistance, a useful distinction to make is between sensor resistances which are large or small in comparison with the connection resistance. In the latter case, a more sophisticated approach is needed.

3.5.1 Thermistor resistance measurement

The case of measuring a resistance which is large compared with the connection resistance is straightforward, as no allowance is needed for the uncertainties associated with the connection resistance. Figure 3.19 shows a circuit to convert an unknown

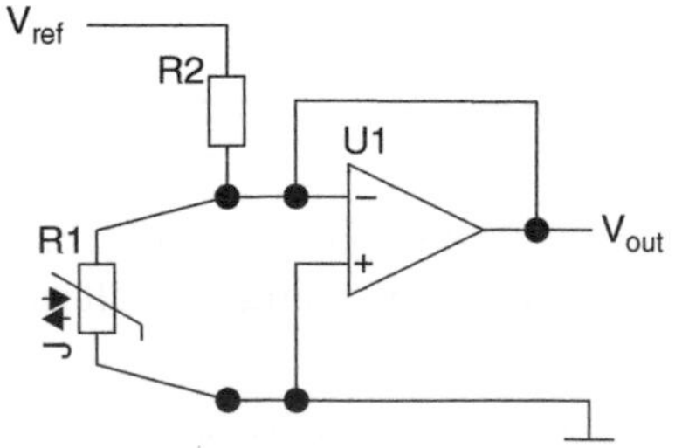

Figure 3.19 Measurement of a varying resistance R1, using a fixed reference voltage V_{ref}, and a fixed reference resistor R2.

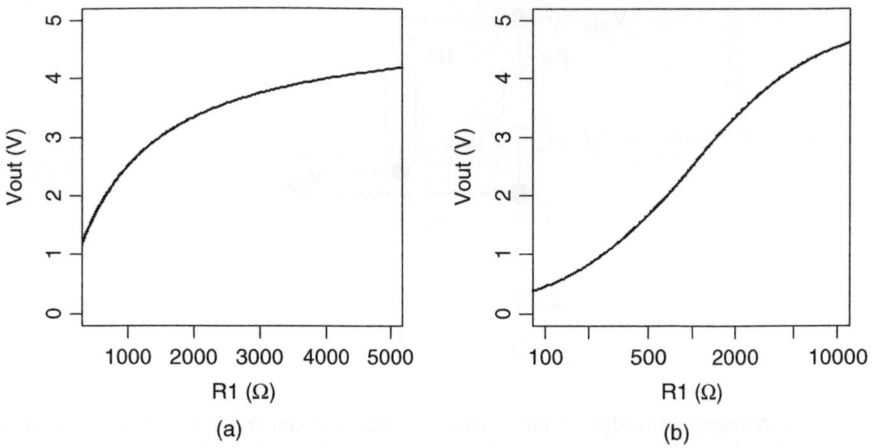

Figure 3.20 Output voltage of the circuit of Figure 3.19 for variations in R1 on (a) linear axes, and (b) a logarithmic horizontal axis, calculated for R2 = 1.0 kΩ and V_{ref} = 5.0 V.

resistance R1 to a voltage for measurement. A fixed reference voltage V_{ref} is required, and a known reference resistance R2. R1 and R2 form a potential divider for the reference voltage, which is buffered by the unity gain opamp stage U1 to avoid loading of the resistors by voltage measuring circuitry. The output voltage is then

$$v_o = V_{\text{ref}} \frac{R_1}{R_1 + R_2}, \tag{3.10}$$

allowing the resistance R1 to be found as

$$R_1 = R_2 \left[\frac{v_o}{V_{\text{ref}} - v_o} \right]. \tag{3.11}$$

Whilst this is not a simple linear response between R1 and v_o, if R2 is chosen to be comparable with R1 at the middle of the measurement range, the response is reasonably linear for small changes in R1 around the middle value (Figure 3.20a). Over a wider range of resistance values, however, the response is approximately logarithmic (Figure 3.20b), which can be useful for linearising the response of some sensors.

An advantage of this circuit configuration (e.g. compared with the linear resistance measurement available from Figure 3.7a) is that it can be used straightforwardly with long connection cables if the resistance to be measured is returned to one of the supplies. For example, the resistance R1 can be mounted at the end of a single-core screened cable, with the outer screen grounded.

3.5.2 *Resistance bridge methods*

The resistance of some sensors in strain gauges which are sometimes used to measure the effect of pressure differences can be quite small, as can that of the resistance

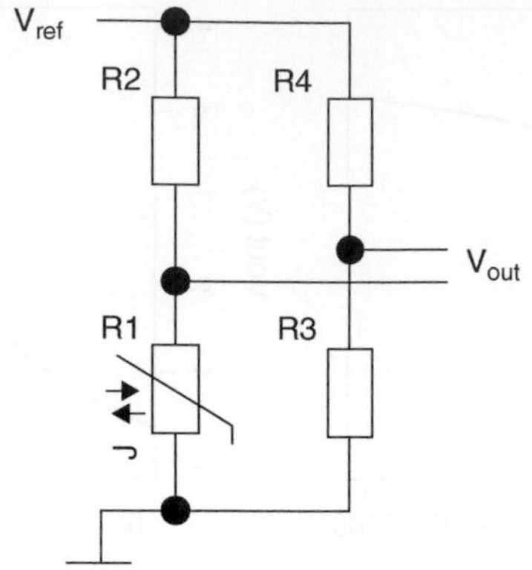

Figure 3.21 A Wheatstone bridge circuit, in which a sensor resistance R1 is to be compared with a range of standard resistances represented as R3, with R2 and R4 also fixed. The output voltage V_{out} is zero – and the bridge is said to be in balance – when the resistance ratios R1/R2 and R3/R4 are equal.

elements used in metal thermometry such as platinum resistance thermometers. If the sensor resistance is smaller or comparable with the connection resistance, appreciable resistance errors can be introduced and a different approach is needed. Some of these are based on the concept of a resistance 'bridge', which is a configuration of resistances allowing resistance *ratios* to be compared. The most well-known resistance bridge circuit is the Wheatstone bridge,[xviii] Figure 3.21, which consists of two vertical arms, each of which are potential dividers for the driving voltage V_{ref}. One arm contains the resistance to be measured (R1); the other three resistances are known. The midpoints of the potential dividers are compared (i.e. the voltage difference v_0 between the R1–R2 and R3–R4 junctions), and the voltage difference is given by

$$v_o = V_{\text{ref}} \left[\frac{R_1}{R_1 + R_2} - \frac{R_3}{R_3 + R_4} \right] . \tag{3.12}$$

In its simplest form, the Wheatstone bridge is adjusted to obtain a balance condition of zero voltage difference, or no current flow, between the R1–R2 and R3–R4 junctions. This occurs if the ratios R1/R2 and R3/R4 are equal, independently of V_{ref}. To achieve the adjustment for the balance condition, a Wheatstone bridge measuring box will provide a wide range of selectable calibrated resistances for R3, and the stable resistances required for R3 and R4.

If, after the balance point is set, the value of R1 changes by a small amount r ($r \ll$ R1), the bridge voltage v_o will also change slightly, varying with r as

$$v_o(r) = V_{\text{ref}} \left[\frac{(R_1 + r)}{(R_1 + r) + R_2} - \frac{R_3}{R_3 + R_4} \right] . \tag{3.13}$$

[xviii] Sir Charles Wheatstone (1802–1875) promoted this circuit configuration, originally devised by Samuel Christie (1784–1865).

The sensitivity of the bridge voltage to the resistance change is

$$\frac{dv_o}{dr} = V_{\mathrm{ref}} \frac{R_2}{(R_1 + r + R_2)^2}, \tag{3.14}$$

or, when r is small,

$$\frac{dv_o}{dr} \approx V_{\mathrm{ref}} \frac{R_2}{(R_1 + R_2)^2}. \tag{3.15}$$

Hence, for small changes in R1, the bridge voltage variation with r is approximately constant, with a sensitivity determined by R1 and R2.

For resistance changes which are appreciable or when there is a finite connection resistance, an alternative is approach is to linearise the bridge using an opamp, as used in a resistance thermometer application [37]. An example is given in Figure 3.22.

The voltage sensitivity of the output to a resistance change r is given by

$$\frac{dv_o}{dr} = V_{\mathrm{ref}} \frac{R5}{R2\ R3}, \tag{3.16}$$

where R2 = R4 and R3 = R1 at some fixed temperature within the range to be measured. A stable reference voltage V_{ref} is required, which, together with the choice of R2, determines the sensor's excitation current, which has to be kept small to minimise self-heating of the sensor. Figure 3.23 shows the response of the circuit to a range of resistances produced using an accurate resistance box; the response is highly linear and is found to be reproducible between different units.

A fuller development of the bridge concept uses different connections for the excitation current and the measurement voltage, to remove the effect of voltage drop

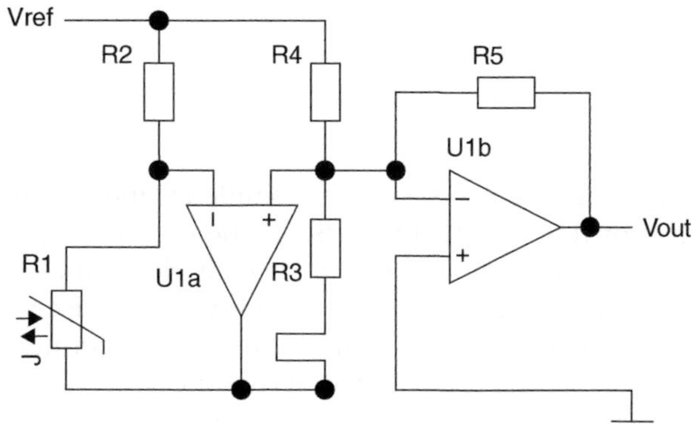

Figure 3.22 Bridge linearisation using an opamp (U1a), including a compensation loop at R3 for the connection resistance to R1. The compensation loop uses an identical length of identical connection wire to that used in connecting R1. The second stage (U1b and R5) is a current to voltage converter.

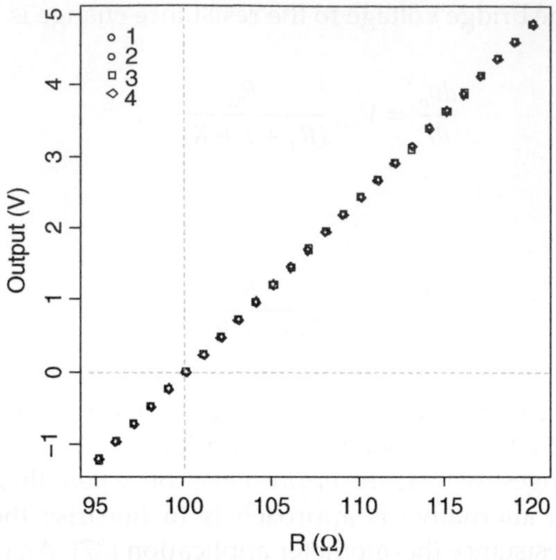

Figure 3.23 Response of the circuit of Figure 3.22 to a range of fixed input resistances *R*, when $V_{ref} = 5.00$ V, R2 = 100.0 Ω, R3 = 10.0 kΩ and R5 = 51.8 kΩ. The plot shows measurements from four different devices, all of which have a very similar, linear output voltage response to the input resistance variation.

due to the excitation current. Separating the sensor excitation current from the sensor voltage measurement leads to a configuration known as a *Kelvin connection.*[xix]

3.6 Oscillatory signals

Signal processing of low frequency signals, such as voltage amplification, can usually be achieved by the opamp circuitry already discussed. But it is also useful to appreciate how a frequency can be generated electronically, as this provides the basis for interval timing as well as a frequency to voltage conversion which is needed for some environmental sensors. A device for generating a stable frequency is known as an *oscillator*.

3.6.1 Oscillators

An oscillator is usually intended to generate a single-frequency signal, defined by some basic properties of its constituent parts. In the case of a mechanical oscillator such as a simple pendulum, the frequency is solely defined by the length of the pendulum. Similarly, in an electronic system, the charging and discharging of a capacitor provides a regular oscillation, which can be used to generate a sine or square wave. However, in order that the oscillation does not die away, some regular energy replenishment is needed, which is usually provided by feedback from the output

[xix] Lord Kelvin (William Thomson, 1824–1907) was one of the greatest Victorian scientists and a president of the Royal Society, who made substantial contributions to thermodynamics and electrical engineering.

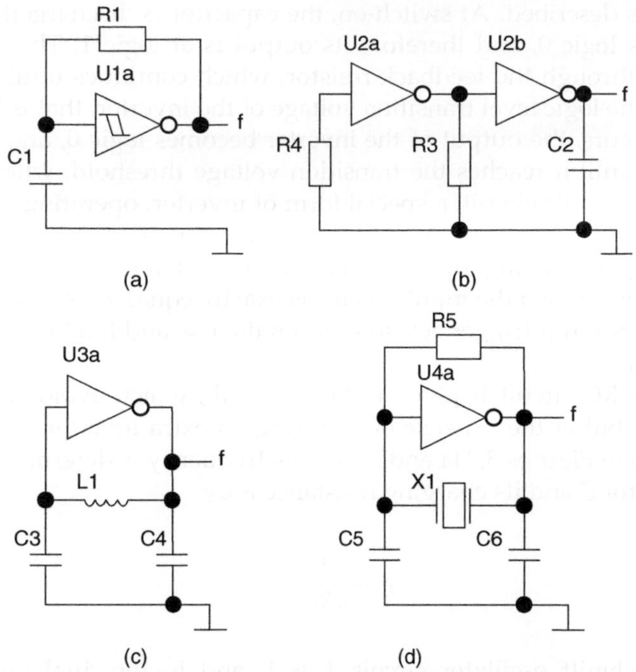

Figure 3.24 Electronic oscillator circuits based on digital inverters. Resistor-Capacitor (RC) oscillators are shown in (a) and (b), a Colpitts inductor-capacitor (LC) oscillator in (c) and a quartz crystal oscillator in (d). The frequency determining components are: (a) R1 and C1, (b), R3 and C2 (c) L1 and (C3 + C4) and (d) the crystal X1. (The inverters are contained within type 40106 and 4069 integrated circuits, with their power supply connections omitted for clarity.)

of the oscillator circuit to the input. To sustain the oscillation through constructive interference, this feedback must be exactly in phase with the original oscillation at the frequency concerned. Many circuits have been implemented to do this, often for radio applications, with the circuit configuration often still known by the name of the originating engineer.[xx] One parameter used to describe an oscillatory circuit is the *Q* (or quality) factor. A 'high *Q*' oscillator is one which decays slowly, and therefore requires little energy to perpetuate the oscillation: it is also one for which the resonant frequency is sharply defined.

A useful circuit component for simple oscillator circuits is the digital inverter (see Section 3.2.1), whose output is defined to be the opposite of its input, in terms of voltages used to represent the two different logic levels 0 and 1. These inverter stages are available in integrated circuits, providing many independent inverters in one package. By driving the input of an inverter from a capacitor charging from the same inverter's output, the output will change every time the capacitor voltage crosses the inverter's transition threshold voltage between logic 0 and 1.

Figure 3.24 shows examples of oscillator circuits based on this principle. In Figure 3.24a, a single resistor–capacitor ('RC') oscillator is implemented using a

[xx] Colpitts or Hartley style oscillators using valves were simple circuits which provided the basis of early radiosonde transmitters.

single inverter as described. At switch-on, the capacitor is discharged, so the input to the inverter is logic 0, and therefore its output is at logic 1. The capacitor will begin to charge through the feedback resistor, which continues until the capacitor voltage reaches the logic level transition voltage of the inverter, that is logic 1. When this transition occurs, the output of the inverter becomes logic 0, and the capacitor then discharges until it reaches the transition voltage threshold, when the process reverses. For this simple circuit, a special form of inverter, operating on the Schmitt trigger principle[xxi] is required, as this has well-defined but different threshold voltages for rising and falling input voltages, to prevent 'indecision' (and associated multiple transitions) when the input voltage is exactly equal to its threshold voltage. Accordingly, the Schmitt trigger acts to sharpen the rise and fall times of an applied digital waveform.

An alternative RC circuit is given in Figure 3.24b, which avoids the need for a Schmitt inverter, but at the expense of requiring an extra inverter. For the two RC oscillator circuits in Figures 3.24a and 3.24b, the frequency is determined by the values of the capacitor C and its charging resistance R by

$$f = \frac{1}{kRC}, \tag{3.17}$$

where, in the Schmitt oscillator circuit, $k \approx 1$, and in the dual inverter circuit, $k \approx 2.2$.

An inductor provides an alternative oscillator component, yielding a more stable frequency than for an RC oscillator. Furthermore, if the positive drift with temperature of the inductor is compensated with negative temperature drift capacitors, the temperature stability of the inductor–capacitor ('LC') oscillator can also be improved over the RC circuit. 3.24c shows an LC oscillator circuit using an inverter based on the Colpitts configuration, in which both capacitors contribute to defining the operation frequency. The oscillation frequency is

$$f = \frac{1}{2\pi\sqrt{LC}}, \tag{3.18}$$

where C is the sum of the capacitances of the two capacitors.

RC and LC circuits are found in instrumentation when the stability of the oscillation frequency is not particularly critical, or when the inductor, resistor or capacitor form a part of a transducer for a parameter to be measured. Examples of this occur in hygrometry (Section 5.6), for sensors which have a capacitance which varies with the relative humidity, and in pressure measurement (Section 7.2.4), where the physical distortion of a pressure capsule can be used to change the geometry of an inductor and, as a result, the associated inductance.If the oscillator frequency is

---

[xxi] This important threshold detector was invented by Otto Schmitt in 1934 for valve circuits. Utilising different transition thresholds for opposite directions of change is known as *hysteresis*; the Schmitt trigger schematic symbol itself includes a hysteresis curve.

required to have very high stability with temperature such as for time measurement, the frequency defining component will usually be a quartz crystal, cut to resonate at the frequency required. Standard frequency crystals (e.g. for frequencies of 1 MHz, 10 MHz and 32,768 Hz, which generates 1 Hz on division), are very inexpensive. The quartz crystal represents a complicated combination of resistance, capacitance and inductance, for which the resonant frequency is both well-defined and stable with a high Q factor. Figure 3.24d shows the circuit for a crystal oscillator using inverters. The capacitors play little role in defining the oscillation frequency, changing the frequency by little more than a few tens of parts per million (ppm) depending on the values chosen. This variation, however, still allows small adjustments in frequency if one of the capacitors is made adjustable, which is known as trimming.

3.6.2 *Phase-locked loops*

Methods for frequency and interval measurements (see Sections 4.1.2 and 4.1.3) assume well-defined pulses of steady amplitude and fast rise and fall times, able to be counted by digital circuitry. Some digital signals, however, may be of poor quality, with variable amplitude and contaminated by noise unrelated to the frequency being measured. An example is the digital signal detected by the receiver in an ultrasonic anemometer (Section 8.2.6), which is ultimately used for timing the transit of an acoustic pulse through air, but may initially have a poor wave shape. For such signals, some signal recovery may be appropriate before the pulses are processed further. The *phase-locked loop* (PLL) provides a highly effective method for doing this.

A PLL uses a controllable oscillator to generate a square wave replica waveform having the same frequency and phase as its input waveform, usually applied through a voltage comparator to obtain a digital signal. Figure 3.25 shows the key elements of a PLL an oscillator, phase detector and filter. The oscillator is voltage controlled, able to generate a range of frequencies between a lower and upper frequency as the control voltage is varied. A phase detector compares the phase of two waveforms of the same frequency, and provides an output voltage proportional to their phase difference. In a PLL, these two waveforms are the input frequency f_{in} and oscillator frequency f_{out}. The phase differences between f_{out} and f_{in} cause the phase detector to generate a phase error voltage, which is used to adjust the oscillator until f_{out} and f_{in} agree in both frequency and phase.

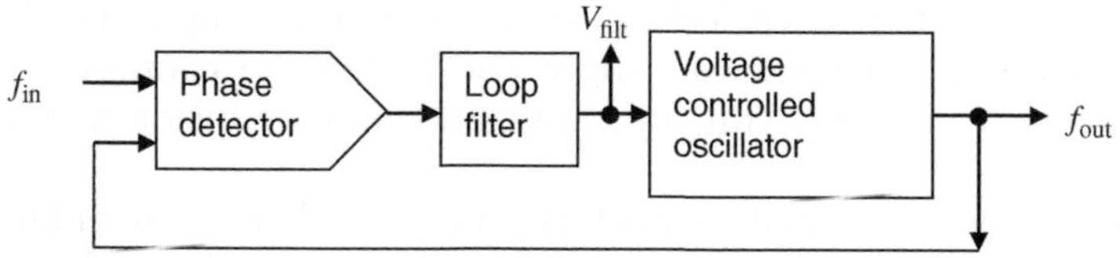

Figure 3.25 The phase locked loop concept. An input frequency f_{in} is compared with a locally generated frequency f_{out} from a voltage-controlled oscillator (VCO), using a phase detector. The output of the phase detector is smoothed with a filter to give V_{filt}, which is then used to adjust the voltage controlled oscillator (VCO) to bring the two frequencies into agreement, in both frequency and phase.

The loop filter defines the overall time response. Generally this is chosen so that many cycles are required before the loop becomes locked. This provides a stabilising effect like that of a flywheel, so that even if occasional pulses are actually missing from the original waveform, a regular square wave will still be provided for subsequent measurement.

Another use of a PLL is in identifying specific frequencies. This requires the adjustment range of the VCO to be restricted, so consequently, only a narrow range of frequencies will allow the loop to lock. This gives a frequency (or tone)-detecting capability, which, if two different tones are chosen, allows digital data transmission, such as for data transfer from radiosondes.

3.6.3 *Frequency to voltage conversion*

The series of pulses generated by some sensors, such as pulse anemometers, may only be able to be measured as a voltage on some recording systems. Conversion of the frequency to a voltage is then required, for which several techniques exist. A simple method is to use the pulses to charge a capacitor, and measure the mean voltage on the capacitor. This approach is sometimes known as a ratemeter circuit. A disadvantage is that the results are closely dependent on the time constants of charging and discharging of the capacitor, which in turn restricts the method to a narrow range of frequencies.

A PLL (Section 3.6.2) can provide frequency to voltage conversion, if the feedback voltage generated to control the oscillator (V_{filt} in Figure 3.25) is measured. This is because, if the relationship between the control voltage and the oscillator frequency is linear, V_{filt} will track the input frequency f_{in} closely to ensure the lock condition of the loop. In order to ensure little loading of the filter a buffer stage is generally needed. The time constant of the loop filter must also be longer than the typical periodic time of the input waveform.

3.7 Physical implementation

If signal processing electronics is installed near sensors at a measurement site, it is likely to encounter a very wide range of atmospheric conditions of temperature, humidity and precipitation. Of these, ingress of water is the most damaging to electronic systems, as it leads to corrosion of circuit connections and damage to circuit boards. Moisture-induced leakage paths will also compromise the reliability of the circuitry. As a result, water-tight enclosures and connectors are essential, but even then water may still somehow find its way inside instrumentation (see Figure 3.26).

Wide ranges of enclosures and connectors giving good environmental protection are available. These are classified in terms of the protection they offer, from water splash resistance to full immersion.[xxii] Long-term durability also requires any polymer components to be resistant to sunlight. Some instruments, such as high gain

[xxii] An 'IP68' classified enclosure is both dust-tight and able to withstand continuous immersion.

Figure 3.26 Damage to an electrometer instrument exposed for many years to Scottish weather. Despite the inner aluminium die-case box, waterproof enclosure and waterproof connectors, small distortions in seals have allowed water to enter the instrument with substantial consequences for its operation.

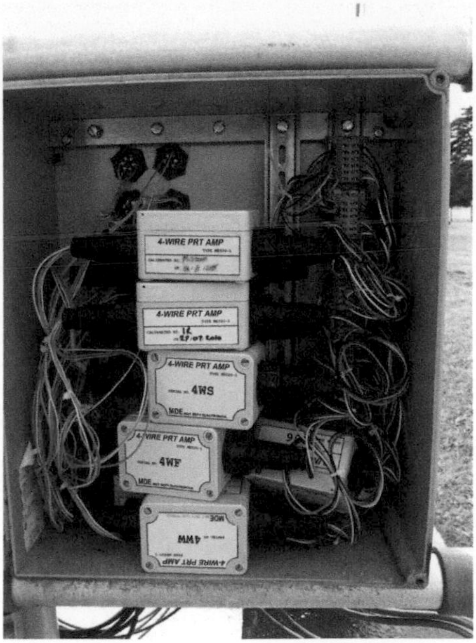

Figure 3.27 A set of platinum resistance thermometer signal conditioning amplifiers used for soil temperature measurements. Each channel's signal conditioning electronics is mounted within an individual weatherproof box, with weatherproof connectors at each end (sensor connections at one end, power supply and signal output at the other). The set of devices is mounted within an outer weatherproof cabinet.

amplifiers, also require electrical screening, and therefore a metal, or metal-loaded box, with suitable seals will be needed to provide good environmental protection. However, a multiple-layer approach to waterproofing is generally advisable, with the signal conditioning circuitry mounted within a weatherproof inner enclosure, protected by a further weatherproof outer enclosure. As cable entry and exit are vulnerable failure points, good quality connectors, mounted correctly with seals and clamps, are also essential (Figure 3.27).

4

Data Acquisition Systems and Initial Data Analysis

Recording sensor variations after applying appropriate signal processing is fundamental to environmental monitoring and measurement. Automatic marking of values using ink on paper charts or mechanical scribing on plates – autographic recording – was an early method, with continuous photographic recording [38] used in environmental science[i] from about the 1860s. This required chemical processing of the photographic paper employed. Electronic recording equipment, such as the UK Met Office's MODLE system (Met Office Data Logging Equipment), was used for recording solar radiation measurements from the 1960s, which used punched paper tape as the storage medium [39]. Both paper and tape records are fragile media with finite working lifetimes. In contrast, digital information is degraded by neither analysis nor retrieval, requiring only good custodianship of the data files to provide indefinite storage.

4.1 Data acquisition

Environmental measurements are presented by sensors in different forms, such as a series of pulses to be counted or time-stamped, or a fluctuating voltage to be sampled. These different data sources are now considered in terms of acquisition and storage for further analysis. Conversion of measured parameters to a digital form is an essential first stage in the data acquisition, as is some method of time-keeping, to mark when the measurements were obtained. Beyond the actual measurements, detailed information about the equipment in use, their calibrations, situation and geographical location – metadata – are also essential requirements, particularly if the data will be processed by third parties not involved with the original measurements. Acquisition and storage from a variety of environmental sensors is most conveniently undertaken using a *data logger*. These measurement and storage systems are designed to be versatile devices, programmable to make measurements in batches or

[i] These techniques were pioneered by Lord Kelvin in his pursuit of characterising atmospheric variability. Indeed, through this technology and related field work in which theories were developed in response to observations, he can be considered to be amongst the first modern environmental physicists.

Meteorological Measurements and Instrumentation, First Edition. R. Giles Harrison.
© 2015 John Wiley & Sons, Ltd. Published 2015 by John Wiley & Sons, Ltd.
Companion website: www.wiley.com/go/harrison/meteorologicalinstruments

Figure 4.1 Photograph of an environmental data logger, deployed within a weatherproof enclosure. A range of environmental sensors measuring wind speed, temperature and radiation has been connected.

continuously and to interrogate the sensors connected in different ways. Figure 4.1 shows a data logger in use. A characteristic feature is a multi-way connector block, to allow the wiring of multiple diverse sensors, and a rugged weatherproof housing.[ii]

At the heart of a data logger is a small computer which runs a logging program to determine what is measured and when, and to apply simple processing to the data. The logger program will be specific to the measurement situation and the sensors in use, and the measurements, or the results of simple processing on the measurements, will be stored in the data logger's memory. Figure 4.2 provides a block diagram of the typical functional components found within a data logger, for data acquisition from multiple-sensor input channels of different kinds.

When more than one sensor is measured, the computer first selects the particular sensor required using a *multiplexer*. A multiplexer is a computer-controlled switch which allows each sensor to be chosen in turn for sampling. Much less circuitry is required for this sequential switching approach than would be the case if all the channels were sampled simultaneously (see Figure 4.4). A necessary requirement for multiplexing is that the switching and sampling of all the channels can be completed more rapidly than the rate at which significant changes occur in any single data channel. In environmental applications, this is usually easily justified, but there may be cases when rapid transient events are sought, such as monitoring responses to lightning strikes, leading to a risk that an event may be missed. In such circumstances, a conditional sampling approach may be useful, to make measurements in response to an event (see also Section 4.1.5).

[ii] The electronic circuitry of a general purpose data logger will be engineered to withstand electrostatic discharge, moisture ingress and connection errors.

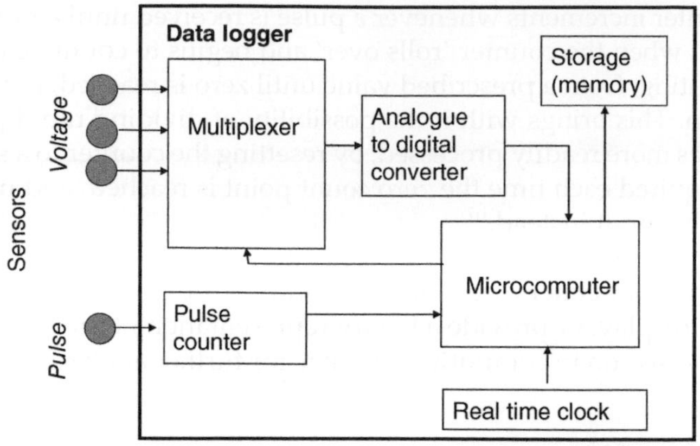

Figure 4.2 Typical components present within a digital data logger used for environmental measurements from multiple sensors providing a mixture of pulse and voltage information.

Data loggers are designed to operate with low power requirements, as they may be required to function indefinitely at a remote site without mains electricity. They will usually include internal batteries to power the microcomputer. Some data loggers have low power (or 'sleep') modes, in which negligible power is consumed until the device goes into a more active sampling mode. The overall power consumption also depends on what is needed to power the sensors, which may be provided from the data logger supply or independently. For measurement systems with more substantial power requirements, solar or wind generated electricity can be used, backed up by a storage battery.

Some form of time or date recording is also usually needed if the measurements obtained are to be compared with other measurements or theory. This necessarily requires the data logger to have its own time and date information, which is provided by the real-time clock.

Aspects of processing of pulse and voltage data are now considered in turn.

4.1.1 *Count data*

Count data originate as a succession of pulses associated with an event, and are recorded to determine how many events occur, the times at which the events occur, or the event rate. An example is a tipping-bucket rain gauge (see Section 10.4.2), which generates a pulse whenever a fixed volume of rain is collected. For such a device, the total quantity of rain in a given interval such as a day may be wanted, or the variation of rain rate with time.

Count data are digital in the sense that they are either present or absent, however the events still need to be captured as they occur, even if they are brief. Recognition of a voltage pulse to be measured requires that it unambiguously exceeds a threshold value. The simplest input circuitry able to achieve this is a voltage comparator or Schmitt trigger (Section 3.2.1), which can compare an input voltage with a reference voltage and provide a triggering pulse if the reference voltage is exceeded. Further processing of the triggering pulse usually requires a digital counter, which is a device (or its equivalent in software) able to store the number of pulses received since it was last reset. A counter increments whenever a pulse is received until a finite maximum value is reached, when the counter 'rolls over' and begins to count again from zero. Downward counting, from a prescribed value until zero is reached, is another possible configuration. This brings with it the possibility of dividing rapid pulse rates to slower pulse rates more readily processed, by resetting the counter to a starting value of the divisor required each time the zero count point is reached, and using the zero transition as at the input instead.[iii]

The value of the counter can be read at fixed intervals, or the time when the counter has reached zero (from either direction) is recorded. Counters may provide a direct digital result for display, or provide a binary representation of the number of pulses received, to be passed on to computing circuitry for further analysis.

4.1.2 *Frequency data*

Frequency data are similar to count data, in that they provide a succession of events to record, but unlike count data, which are in general irregular, frequency information is periodic. An example of an instrument able to generate a frequency is a cup anemometer, which produces a regular pulse stream as it rotates at constant speed.

The input voltages will be presented to a voltage comparator (Section 3.6.1) for the pulse train to be clearly defined with accurate rising and falling edges for the following circuitry. Measurement of a frequency is a process similar to that for processing count data, except that the individual pulses contained will not be time-stamped. In a frequency counter, the pulses will be counted in a known period, and the total number of pulses determined. The period for which the pulses are counted is sometimes known as the *gate time*. Because a pulse at the beginning or end of the gate time may only be partially completed, there will be, on average, an uncertainty of ±1 in the total number of pulses measured during the gate interval. For a gate time T_g during which N pulses are counted, the pulse frequency f is given by

$$f = \frac{N}{T_g}. \tag{4.1}$$

If the gate time is short, the ±1 pulse count ambiguity may become a large factor in the uncertainty of the final frequency result. However, if the gate time is long, it will not be possible to determine the frequency in a reasonable time, or to follow slow changes in frequency. For low count rates, one alternative is to obtain the frequency using reciprocal counting, that is by determining the mean interval between pulses, and finding the frequency from the reciprocal of the mean interval. Another possibility is to convert the frequency to a voltage (Section 3.6.3) and record the voltage instead.

4.1.3 *Interval data*

Measurements of intervals can be achieved using well-defined timing pulses, which are usually derived from a quartz oscillator circuit (e.g. Figure 3.24d). To measure an

[iii] Frequency division is also sometimes known as *prescaling*.

interval, the timing pulses occurring between one event and the next (or a start and stop signal) are counted. If the timing oscillator frequency f_{osc} is known, the interval T between events when N pulses are counted is simply

$$T = \frac{N}{f_{osc}}. \tag{4.2}$$

As for the case of frequency data, a timing oscillator pulse could overlap with an event pulse, hence, on average, the number of pulses is only reliably known to ± 1. As long as N is large (i.e. obtained over a long interval with a high frequency timing oscillator), the related uncertainty will be negligible. Should the pulse rate be comparable with the timing oscillator frequency, however, a substantial uncertainty in the interval determined may result. A higher frequency timing oscillator is then required.

4.1.4 *Voltage data*

A varying sensor voltage is a continuous variable (in the sense that its instantaneous value can be any real number), and is continuous in time (i.e. it has a value at all times, whether or not it is actually being sampled). Accurate conversion of continuously varying electrical signal into a digital signal therefore forms a central aspect for a data logger. The process of converting an instrument's output voltage to a digital value is known as *analogue to digital conversion* (or digitisation), and uses an analogue-to-digital converter (an 'AtoD' or 'ADC').[iv] The simplest ADC is a voltage comparator (Section 3.2.1), as it can determine whether an input voltage is larger or smaller than a given reference voltage, returning a binary result of 0 or 1. Variations on this simple comparison process can improve the resolution from a single 0 or 1 to a binary number extending to very many digits.

Many well-refined ADC methods are available, usually all contained within a single electronic component. An obvious extension of the single comparator is to use a set of multiple comparators, each supplied with an increasing sequence of known reference voltages, but with the same input voltage presented to them all. By knowing which comparators trigger (i.e. have reference voltages less than or including the input voltage), the input voltage can be determined as lying between the greatest reference voltage applied to a voltage comparator which has not triggered and the immediately adjacent comparator which has triggered. Whilst this method requires many comparators if good resolution is to be obtained, it has the advantage that, as all the comparators operate in parallel, it is very fast, for which reason it is also known as 'flash conversion'. Another approach is to use only one comparator, but vary the reference voltage until the comparator indicates it is very close to the input voltage. If the reference voltage can be controlled digitally with a high resolution (for example using a switched potential divider, known as 'successive approximation', or the timed charging of a capacitor, known as 'integrating' or 'slope' conversion), comparable resolution becomes available for recording the input voltage. These outline

[iv] A digital to analogue converter (or 'DAC') performs the inverse process of an ADC: by sending a DAC a digital code, a unique voltage can be generated.

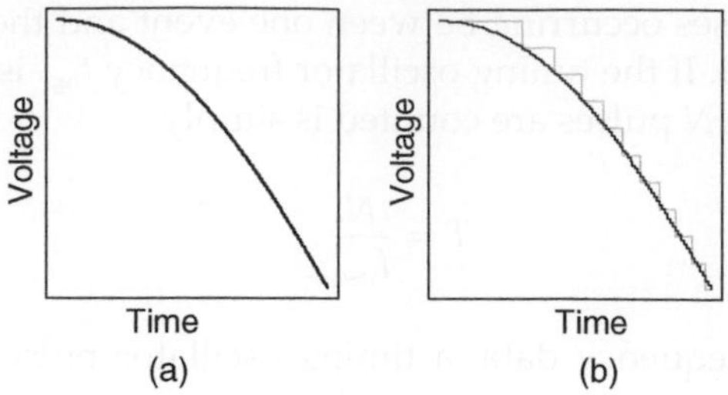

Figure 4.3 Effect of ADC resolution on digitisation. (a) shows a changing voltage to be digitised. (b) shows the representation of the voltage in (a) using coarse resolution (grey line) and at high resolution (black line).

examples also indicate a general principle, that higher resolution conversion requires a longer time than lower resolution conversion, and/or more extensive circuitry.

The outcome of the analogue-to-digital conversion is a numerical result expressed in the binary number system. For binary numbers, each digit is known as a bit, and hence increasing the number of bits in a binary number increases the size of the number. The maximum value of a binary number containing n-bits is 2^n-1, as its lowest value (all n bits zero) is zero. Consequently, an ADC with a larger number of bits available to represent a voltage will do so with greater resolution than for an ADC using fewer bits. The effect of finite ADC resolution is therefore to round the input voltage to the nearest numerical value available (Figure 4.3).

Table 4.1 compares the voltage resolution obtained by ADC operating with different number of bits.

The voltage resolution also depends on the range (or span) of voltages to be digitised, as, for any particular ADC, only a fixed number of values is available depending on the number of bits it uses. The voltage resolution ΔV for a voltage span V_{span} is

$$\Delta V = \frac{V_{\text{span}}}{2^n - 1}, \tag{4.3}$$

where n is the number of bits available to the ADC. Hence if the span of voltages to be sampled can be restricted, the voltage resolution can be improved, for example if only a 1 V unipolar input voltage range is required, the full resolution of a 12-bit ADC

Table 4.1 Comparison of the voltage resolution obtained from an analogue to digital converter (ADC) of different bit resolutions, for a unipolar 5 V full-scale voltage

Number n of bits used by ADC	Maximum value (2^n-1)	Voltage resolution for 5 V full scale
8	255	19.6 mV
10	1023	4.9 mV
12	4095	1.2 mV
16	65,535	76.3 μV
20	1,048,575	4.8 μV

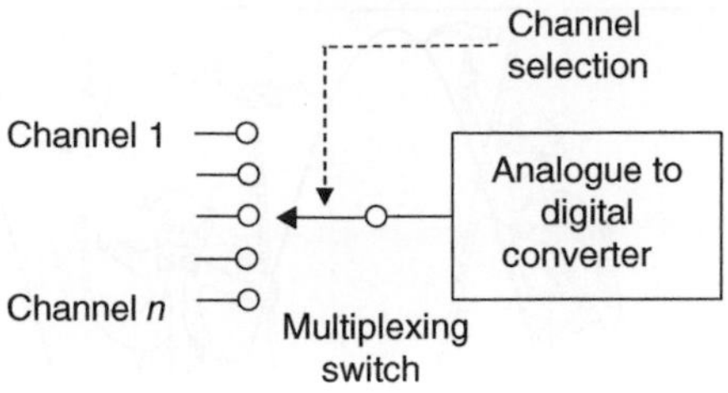

Figure 4.4 Principle of multiplexing, in which the input of the analogue to digital converter is switched, in turn, to measure each of n input channels.

will be about 0.25 mV, but for a ±5 V range (i.e. 10 V span), the resolution will become about 2.5 mV.

An ADC is rarely used with only a single sensor channel, as usually different variables will be compared. Multiple channel operation can be achieved by switching the ADC between sensors, using an electronic multiplexer switch to select the channel required. In this way, a good quality ADC can be deployed on measuring very many different sensors. It is necessary, of course, for the knowledge of which channel is sampled to be recorded in addition, and for the multiplexing rate (the rate at which the ADC returns to the same channel), to be sufficient to record the fluctuations required in the quantity concerned (see also Section 4.1.5.) Figure 4.4 illustrates the principle.

If an ADC's input is not connected to a sensor (i.e. it is left 'open circuit'), the ADC value obtained may fluctuate randomly, or show some relationship with other adjacent ADC channels if used with a multiplexer. This is generally undesirable, as it makes it difficult to recognise that a sensor has become disconnected. One approach to avoid this is to connect all ADC channels to a reference voltage through a high value resistance.[v] This will normally not affect the ADC input voltage provided by signal conditioning electronics, but, if the sensor becomes disconnected from the channel, the ADC value returned will then become a known, fixed value, which can be detected in software.

4.1.5 Sampling

Conversion of a continuous variable to a digital form can only approximate the original continuous signal, as its amplitude variations can only be described using a fixed set of values. The continuous voltage variations occurring therefore become quantised. There is a further consideration, however, in that a signal also needs to be sampled sufficiently rapidly for its significant temporal variations to be recorded. Because an ADC usually takes longer to make a high resolution voltage measurement than a lower resolution measurement, the temporal resolution and voltage resolution may be related, requiring a compromise between the sampling rate and the amplitude resolution. Figure 4.5 characterises both aspects of the sampling process.

[v] This can be known as a pull-up (or, if appropriate, pull-down) resistor.

Figure 4.5 Schematic depiction of the analogue to digital conversion process, for a sinusoidal input signal generated in different environmental science scenarios. The finite separation needed in time between subsequent samples is apparent from the positioning of the measuring operatives, and quantisation of the signal amplitude results from their regularly marked measuring sticks. Both aspects limit accurate representation of the input waveform. (Image by Philip Mills.)

For a voltage which is sampled regularly, there is a maximum frequency which can be determined in the data obtained. This value, the *Nyquist frequency* f_N, is given by

$$f_N = \frac{1}{2} f_s , \tag{4.4}$$

where f_s is the sampling frequency. This arises from considering the sampling of a sinusoidal waveform, and, in particular, how many regular samples would be needed to determine the sine wave's frequency unambiguously. The sampling frequency required and the signal frequency are related, as illustrated in Figure 4.6 for regular sampling of sine waves of different frequencies.

For a sinusoidal waveform of frequency $f = 0.75$ Hz (Figure 4.6a), sampling at once per second yields values which suggest that a lower frequency sine wave of 0.25 Hz is present. For a lower frequency sine wave with $f = 0.6$ Hz (Figure 4.6b), the same sampling rate suggests that a sine wave of 0.4 Hz is present. Only for $f = 0.5$ Hz can the original sine wave be correctly reconstructed from the 1-second sampling (Figure 4.6c). In the first two cases, spurious lower frequency signals are generated by the sampling process, as part of a series of n alias frequencies f' given by

$$f' = 2nf_N \pm f . \tag{4.5}$$

These sampling considerations indicate that, if frequencies greater than $\frac{1}{2} f_s$ are present in the input waveform, aliasing effects may occur. These contributions can be substantially reduced by using a filter to remove the higher frequencies. Such an 'anti-alias' filter is a low pass filter designed to only allow signal frequencies below $\frac{1}{2} f_s$ to reach the ADC.

An approach to improve the resolution of an ADC is to make a rapid burst of N repeated samples of the input waveform, the results of which are then summed. The effective resolution is increased over that of an individual sample by N. This is known

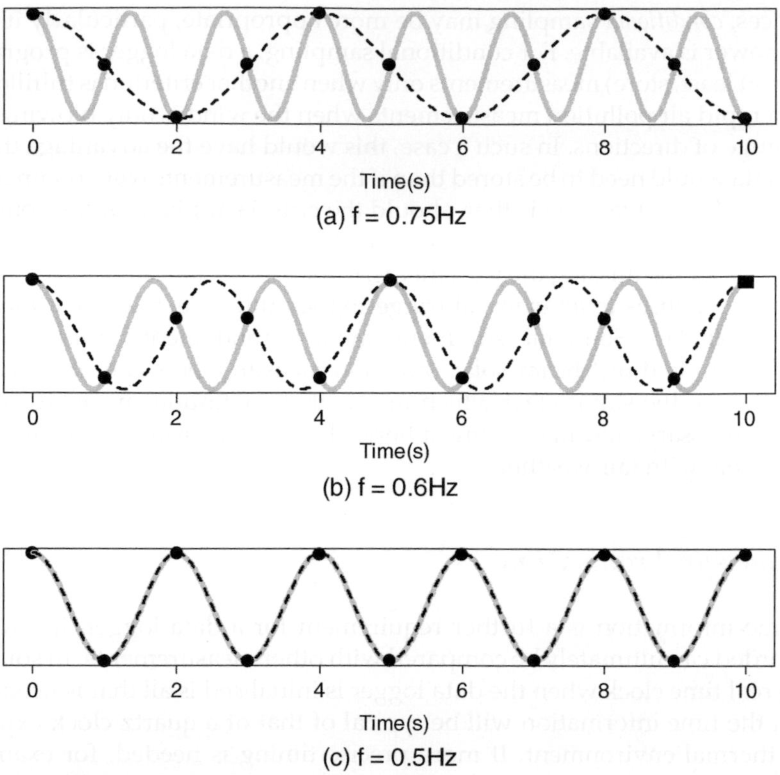

Figure 4.6 The effect of regular sampling at once per second of a sinusoidal waveform (grey line) of frequencies (a) 0.75 Hz, (b) 0.6 Hz and (c) 0.5 Hz. The waveform is then reconstructed from the point samples (dashed lines). The reconstructed waveforms have frequencies (a) 0.25 Hz, (b) 0.4 Hz and (c) 0.5 Hz.

as *oversampling*, and acts to reduce the quantisation error (ΔV in Equation 4.3), due to the limitation of an ADC in having to represent an input voltage as a fixed value when, in reality, the actual voltage will lie between two adjacent ADC values. By making multiple samples, the series of ADC results will contain both values, with their relative abundance in proportion to how far the input voltage lies between the two values. Summing such oversampled ADC results therefore provides a higher resolution result without the need for a higher resolution ADC, but with the associated disadvantage that the conversion time is increased because of the repeated measurements required. In addition, the oversampling will raise the Nyquist frequency, which can improve the effect of an anti-alias filter by moving further into the stop-band region of the filter.

Sampling of an environmental sensor is usually made regular in time for convenience, but there is no reason why this has to be the case. For non-regular sampling,[vi] the Nyquist frequency is found from the mean sampling frequency [40]. In other

[vi] Irregular sampling may lead to less ambiguity in the reconstruction of the sampled waveform than regular sampling and hence the mean f_N can be greater than the local value of $\frac{1}{2}f_s$.

circumstances, *conditional* sampling may be more appropriate, particularly if limited storage or power is available. For conditional sampling, a data logger is programmed to make (or, at least, store) measurements only when another criterion is fulfilled, such as to obtain rapid air pollution measurements when the wind is only blowing from a specified range of directions. In such a case, this would have the advantage that very much less data would need to be stored than if the measurements were retained for all the directions. A disadvantage is that, should the criteria applied for the conditional sampling subsequently appear to be insufficient in some way, there is nothing that can be done about the missing measurements. Conditional sampling is well suited to investigating responses to an external triggering stimulus, such as the atmospheric effects of a solar flare. Some measurements are also fundamentally intermittent in nature, for example direct beam solar radiation measurements can obviously only be obtained when the sky is clear which means that continuous measurements are probably unnecessary, and many atmospheric electricity measurements are usually associated solely with fair weather.

4.1.6 *Time synchronisation*

Reliable time information is a further requirement for a data logger, in order that events recorded can ultimately be compared with other measurements. In some cases setting the real time clock when the data logger is initialised is all that is needed, and the drift in the time information will be typical of that of a quartz clock exposed to a varying thermal environment. If more precise timing is needed, for example, to uniquely identify individual lightning events, other sources of time information are available which can be decoded. The most accurate method is to use time information supplied by the Global Positioning System (GPS) satellites, through decoding the data provided by a GPS radio receiver. A lower resolution method is to use long wave standard time transmissions, such as those sent from Frankfurt or Anthorn for radio clocks. These transmitters provide 1-second time pulses synchronised to Universal Coordinated Time (UTC), which can be received over much of Europe. The time pulses are shortened and lengthened to encode the time and date in a binary format, which can be retrieved from analysing 1 minute of pulses. This method of synchronisation is widely used for clocks and watches, and uses much less power than a GPS receiver, particularly if it is only used to adjust a data logger's internal clock infrequently, such as daily.

Alternatively, the time can be uploaded to the data logger by a remote computer, when the data transfer occurs.

4.2 Custom data logging systems

The data logger principles described in Section 4.1 apply not only to general purpose environmental data loggers but also to a range of computing systems now available such as desktop computers, and miniature computers such as the Arduino or Raspberry Pi. These allow custom data logging systems to be constructed, optimised for fast data acquisition, number of logging channels, storage, size, power consumption or cost.

Figure 4.7 Multiple-channel data acquisition system used at Reading University Atmospheric Observatory, based on data acquisition cards connected to a standard desktop PC, which is also networked for web access to the data. The patchboard of wires allows the channels to be configured to the instruments connected.

4.2.1 *Data acquisition cards*

Standard desktop computers can be expanded into data acquisition systems by the addition of data acquisition cards in expansion slots within the computer, or via standard USB (universal serial bus) interface connections. These cards allow voltage or pulse measurements at considerable rates as they can be configured to access fast memory available on the computer.[vii] Because of the processor time required, it is likely that the computer will need to be dedicated to the task if the data acquisition is to be continuous, with networked transfer of the data files generated at suitable points in the measurement cycle. The use of multiple data acquisition cards makes very many channels available (Figure 4.7), which are sampled and recorded in a consistent manner. A consideration may be the effect of operating system of the computer used, and whether it allows real-time information to be exchanged from the data acquisition system, as sometimes other internal processes (e.g. hard disk access for other software) may be given a priority over storing incoming information.

4.2.2 *Microcontroller systems*

One particular class of computing chip, the micro-controller, is very well suited to data logging. These are small programmable devices which can store a program of a few dozen lines with limited storage. Their main use is as adaptable hardware to format measurements from sensors, for transfer to storage or display devices for other data transfer applications. Because they contain key functional parts such as

[vii] A key consideration is ensuring that reliable software drivers compatible with the computer's operating system exist and will continue to do so, or are available open source to allow modifications and improvements to be made.

multi-channel ADCs, oscillators, timers and counters, they can provide the core hardware for data logging too. Some microcontrollers have been developed to generate inexpensive modular systems, such as the Arduino series of circuit boards[viii] and units, which are programmed in a high level computing language such as C. (The Raspberry Pi is another example of a small computer designed for easy programming which is suitable for some data logging applications.) A related advantage can be that their open source approach to software distribution simplifies the construction of a data logging system for a specific application. Such systems can be made stand-alone, or operated alongside a more substantial computer to provide analysis and storage.

The simplest microcontrollers consume very low power (milliwatts) in comparison with more substantial computing hardware and modules associated with desktop and laptop computers, but have limited arithmetic capabilities. This is because they use the lowest level manipulation of numbers – integer arithmetic with positive integers only – which means that the precision of floating point calculations is not available, or a major additional overhead of programming would be required consuming program space. As ADCs only provide integer values from a digital conversion, the integer arithmetic limitation only becomes a hindrance in data processing, rather than at the acquisition and storage stages. Data manipulation which is efficient in terms of operations or storage, such as the in-place sorting of values within a single data array may also be helpful. With this approach, repeated sampled values could be sorted to allow the median and interquartile range (IQR) to be found. This would allow the spread in a set of integer measurements to be found without the computational complexity of floating point calculations. Even for simple processing however, such as that required for a thermometer to report values in Celsius, floating point representation and negative values would typically be needed, although it can sometimes be elegantly circumvented.[ix]

Obtaining data from additional science sensors carried by radiosondes provides an example of a custom data acquisition application unsuitable for a general purpose data logger. This is because the logger has to be inexpensive as it has a lifetime of only a few hours at most (after which it is almost always lost), it must also be lightweight because of payload considerations, and it must consume negligible power if the radiosonde's radio transmissions and primary meteorological measurements are not to be affected. In one implementation, these requirements have been fulfilled using a microcontroller to manage the data acquisition and transfer to the radiosonde [41], at a current consumption of 3 mA (see Figure 4.8).

4.2.3 Automatic Weather Stations

An integrated data logging system supplied with a power supply and series of sensors specifically intended for meteorological monitoring, such as making and storing frequent measurements of temperature, rainfall, wind, pressure and solar radiation, is also known as an Automatic Weather Station (AWS). These devices operate using

---

[viii] Arduino peripheral circuit boards are also known as *shields* and the Arduino program a *sketch*.

[ix] Interestingly, the otherwise obsolete (at least in Europe) Fahrenheit scale allows basic temperature registration between −17°C and 37°C, at a resolution equivalent to ~0.5°C, using just two positive integer values.

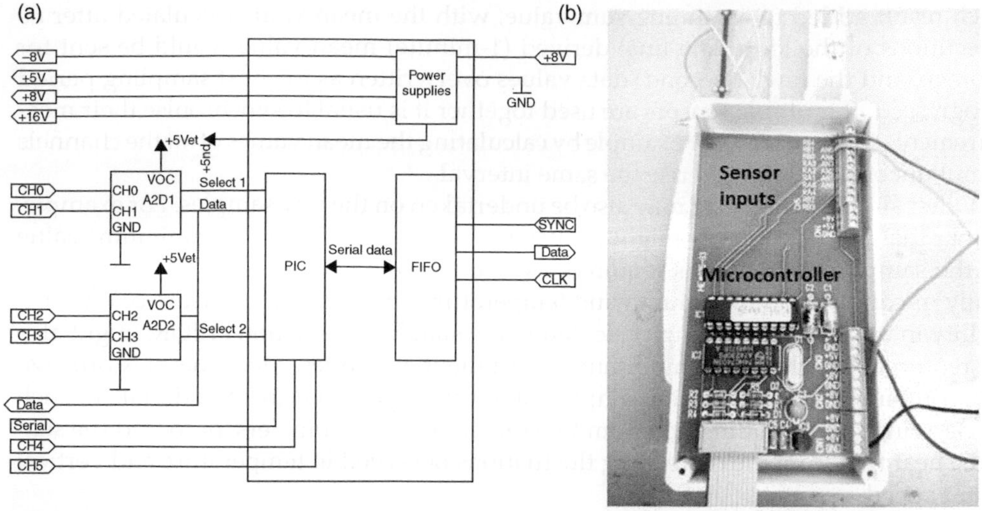

Figure 4.8 Customised 'Pandora' data acquisition system ((a) outline schematic and (b) system view), for a Vaisala RS92 radiosonde, to allow the radiosonde to carry additional science sensors. Data acquisition is managed by a programmed PIC microcontroller, in responses to data requests made by the radiosonde (using the SYNC, DATA and CLOCK lines), using a First-In-First-Out (FIFO) shift register to hold the data briefly for synchronisation with the radiosonde. Analogue input voltages are applied on channels CH0, CH1, CH2 and CH3 via 16-bit analogue to digital converters and low resolution (10 bit) voltage conversions on channels CH4 and CH5. Regulated power supplies for the sensors are also provided (a). (Reproduced with permission of The American Institute of Physics.)

a multi-channel data logger, with multiple sensors and, sometimes, wireless connections; they vary, however, considerably in their cost and quality, both in terms of the capability of their software, and the accuracy and durability of the sensors supplied.

4.3 Management of data files

For all but the simplest application of repeatedly sampling a single voltage channel at a default rate, a data logger will need to be programmed to define the sampling strategy required. The values obtained and stored will also have to be retrieved from the data logger, and copied to a computer for analysis or final storage.

4.3.1 *Data logger programming*

In programming a data logger, choices have to be made about sampling rate and resolution, which depend on estimates of the typical range of values expected and the rates at which important processes or events occur or the response time of the sensor concerned. Some processing may also be required. For example, it may be that a sensor's mean value is only required to be logged every minute, but, for this mean value to be representative, individual samples are needed every second. Programming languages vary between data loggers, but such a measurement strategy could be obtained with little storage using a loop to execute a read of the sensor with each result added to a running sum value, with the mean value calculated after 60 executions of the loop. The final derived (1-minute) mean value would be sent for storage, and the raw (1-second) data values overwritten as the next sampling period progressed. If multiple sensors are used together it is usual to synchronise their measurements in some way, for example by calculating the mean values of all the channels simultaneously to characterise the same interval.

Other simple processing may also be undertaken on the raw samples. For example, rather than mean values, it might be that an extreme (maximum or minimum) value in the sampling period was required, as in the case of determining a wind gust or daily maximum and minimum wind temperatures. A statistical measure of the variability in the samples, through calculating the standard deviation or IQR, might also be required, which can be important for turbulent quantities. More complex processing combining more than one sampled quantity might need to be undertaken, such as deriving the wind direction from two directional anemometers, or to find the sensible heat flux from the correlating fluctuations observed in temperature and vertical wind speed.

Although it is always preferable to store as much raw data as possible in case the processing needs to be modified, it may be more important in some situations to maintain the continuity of measurements. This amounts to establishing a balance between the storage of raw data and the amount of derived data, which is usually determined by how much storage is available, itself decided by how often the data logger's memory can be downloaded and how much can be downloaded at one time. Data transfer by downloading is a key part of operating a data logger as, if the data values are not downloaded and the data logger memory becomes full, either new data will begin to overwrite the values stored or no new values will be obtained. In both cases data values will be lost.

4.3.2 *Data transfer*

Transfer of data from a data logger to a computer can be achieved in several different ways. The most basic method is for a laptop or tablet computers to be taken to the data logger site and the data downloaded via a physical or short-range wireless connection to the logger. This is a particularly appropriate method if the site is accessible and the instruments already require regular checking for other reasons. An alternative method, again if the site is accessible, is for the data storage card to be removed from the data logger and replaced with an empty one, with the full card taken away to provide the data for analysis. If the site is not accessible, a radio modem method can be used. This is achieved by installing a mobile phone subsystem in the data logger, and the mobile phone network is used to retrieve the data. Of course there must be adequate network coverage at the site concerned, and sufficient power to operate the mobile phone. As for taking a computer to the remote site, time and date synchronisation of the logger can also be undertaken, but the amount of data transferred will be limited, both because of the mobile phone network data rates, and the power limitations for the mobile phone transmitter.

For some data logging systems, or for the connection between a computer and a radio modem, the protocol for the data transfer with the computer may have to be precisely specified. There are many standards for serial transfer, which is the method by which data are sent successively as a series of data bits along a single cable or data channel. The most established serial interface, which dates back to teletype machines, sends data between two devices at agreed standardised data rates (given in bits per second, or baud rates). This method[x] is usually very reliable, but only after it has been set up correctly. There are several opportunities for inconsistency as the baud rates must be consistent between the devices, together with any internal checking (parity) bits, start or stop bits and the data polarity (true or inverted). Modern Universal Serial Bus (USB) interfaces use more complicated hardware and software to be largely self-configuring if the appropriate software driver is available, and are able to provide power from the host computer for powering or recharging the data logger.

4.3.3 *Data file considerations*

The organisation of the data file generated by a data logger is essentially decided by the logging program controlling it. Typically, the file will consist of multiple lines of text data,[xi] with each line containing simultaneous data from all the sensors connected. Each line will therefore need to include timestamp information (or at least an index number from which the date and time can be calculated), with the set of values derived from the sensor samples. The individual values in each line will usually be separated by a standard non-numeric character, such as a comma or tab. An alternative to a text data file is a binary data file, in which the values are stored in the form obtained by the ADC directly, and not converted to text representation. These are more compact than text files, but require detailed knowledge of the data format to reconstruct the numeric values stored.

When the data values are downloaded from the data logger, the data may be saved into separate files (e.g. by day or hour), or kept in one large file. Each method has its benefits. In the first case, the risk from corruption of the whole file during transfer (and therefore potentially loss of all the data) is minimised, and if different calibrations are required at different times, these can be applied to the files concerned relatively easily. However, the subsequent analysis of final calibrated values will usually be simpler if a single large file is constructed.

After downloading the data from the data logger, the values obtained will usually be raw values of voltages or pulse rates, and hence calibrations must be applied to convert these to physically meaningful units. These calibrations will either be supplied by the manufacturer, if the sensors were commercially produced, or from previous laboratory tests for any specialist sensors deployed. Following calibration and inspection, erroneous or rogue values obtained should be marked in some way (e.g. by using an exceptional numeric value which can never occur during normal operation, or, preferably, by indicating that the quantity is temporarily 'Not Available', NA as the software permits) so that no further processing of that value is possible. It is generally more useful to retain the structure of a file in terms of regular lines of data despite the presence of multiple instances of NA values, rather than to omit

---

[x] Basic serial interfaces are sometimes referred to by the voltage protocols used to signal the binary data transferred, such as RS232 (which uses bipolar voltages), or RS485 (which achieves a greater range by using an anti-phase drive system).

[xi] Text files are also sometimes known as ASCII files, from the American Standard Code for Information Interchange, which specifies how individual text characters are stored.

Table 4.2 Stages of data processing required to convert data logger values to scientifically-useful values for analysis

Data processing stage	Data contents
Level 0	Raw data logger values
Level 1	Calibrated values
Level 2	Corrected calibrated values and or derived quantities

periods from the raw files when there are no valid values. This fixed number of lines can often simplify subsequent analysis. Table 4.2 summarises the different stages of data processing, from raw logger values to final corrected and calibrated values.

Storage of date and time information is surprisingly troublesome, because of the variety of cultural conventions used in writing the date and time to separate days and months or hours and seconds. A straightforward method is to use a decimal method of recording time, such as decimal hours, decimal year days,[xii] or decimal years. Such a continuously varying decimal time format offers efficient storage in the data file[xiii] which can always be converted back to another date or time format, but, most importantly, it can be plotted simply in spreadsheet software as a time axis without further processing.

4.4 Preliminary data examination

The initial stages of data analysis can be summarised as follows. Data files associated with environmental sensors used with data loggers present raw voltages or count rates (i.e. 'Level 0' data in the terms of Table 4.2), to which a calibration must first be applied to retrieve physical quantities. The next stage of data analysis is to check that the quantities obtained lie within the range of the values expected, and that obvious outliers, such as values which are clearly unphysical for the quantity concerned, are removed. This can be achieved by visual inspection of the data, comparison with similar measurements, data processing to reveal expected features, and comparison with values from theoretical or numerical model values. It is essential, however, to retain the original data, should, subsequently, different corrections and calibrations ultimately be required.

4.4.1 *In situ calibration*

If there is no calibration available to convert the sensor voltages to physical quantities, a further calibration experiment will be needed. For environmental sensors, the comparison with a calibrated (or absolute) instrument can often be made implicitly whilst

---

[xii] Note that, to avoid confusion with astronomical conventions, in which the Julian day number 0 is assigned to the day starting at noon on January 1, 4713 BC (the beginning of the proleptic Julian calendar), the calendar day of the year (i.e. a number from 1 to 365 or 366) is best referred to simply as the *year day*.

[xiii] Consider a measurement made on 31 December 2000 at 1100. This could be stored as 31,12,2000, and 1100, which would require 16 characters, including separators. Alternatively, again resolving the time to the nearest minute, a decimal year format would yield 2000.998520, requiring only 12 characters of storage. Preventing repetition of the year by using annual data files, each containing decimal year days, would be still more efficient.

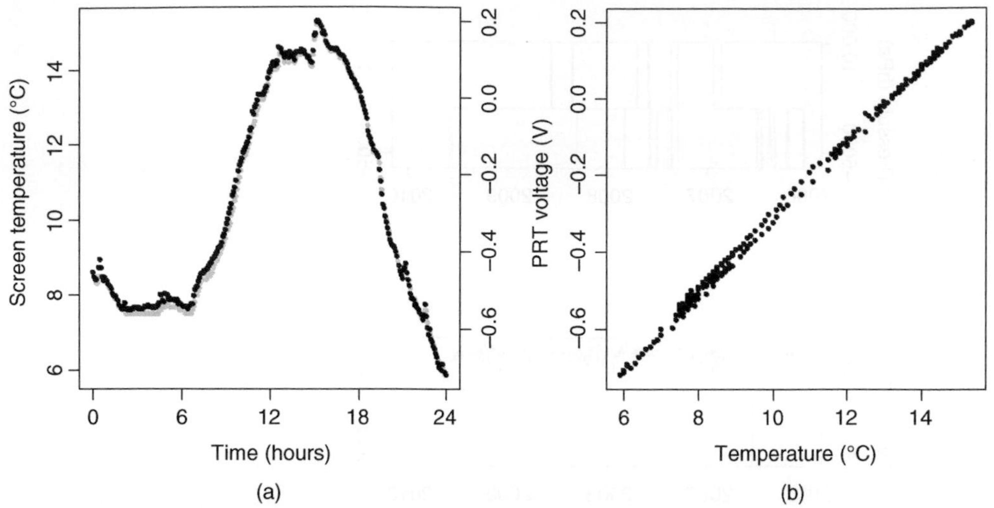

Figure 4.9 Results of a calibration experiment during which a platinum resistance thermometer (PRT) was left inside a Stevenson screen for a day, and the screen temperature and PRT voltage were recorded at 5-minute intervals. (a) shows the simultaneous air temperature (grey points, left-hand axis) and PRT voltage (black points, right-hand axis). (b) shows the PRT voltage plotted against the air temperature. A fitted line to (b) has gradient (0.0970 ± 0.0002) V°C^{-1} and intercept (-1.2974 ± 0.0022) V.

both sensors are exposed to the range of natural values they are likely to encounter in operation, and both sets of measurements are recorded using a digital data logger.

Figure 4.9 shows an example of such a calibration for a platinum resistance thermometer, placed in a Stevenson screen where reference air temperatures are also available from a calibrated thermometer. The two thermometers are then exposed to the diurnal variation in temperature, which is sufficient to allow a linear calibration to be obtained. The calibration is only appropriate over the range of values in the calibration (unless theory indicates a linear response can be expected over a wider range), but this approach has the advantage that, at least for immediate use, the range of temperatures of the calibration is likely to be close to those encountered.

4.4.2 *Time series*

When calibrated values are available as a sequence of regular values with time, basic checks can be made on the range and likely time response apparent in the data. This can be carried out using spreadsheet programs or data analysis packages. Figure 4.10 shows visual checks on data which can be made rapidly, by plotting the values against time (a), and by generating a histogram or frequency distribution from the values (b). In Figure 4.10a, large steps in the data are immediately apparent, emphasised by the line plotting, and the histogram (b) shows the presence of large positive and negative values well away from the central cluster of values. In fact, both the positive and negative extreme values (of −6999 and 9999 respectively) were merely generated by the data logging software, as values used to indicate measurements which were Not Available (NA). The NA values have no physical meaning; they cannot be relevant to the analysis and must not be processed further.

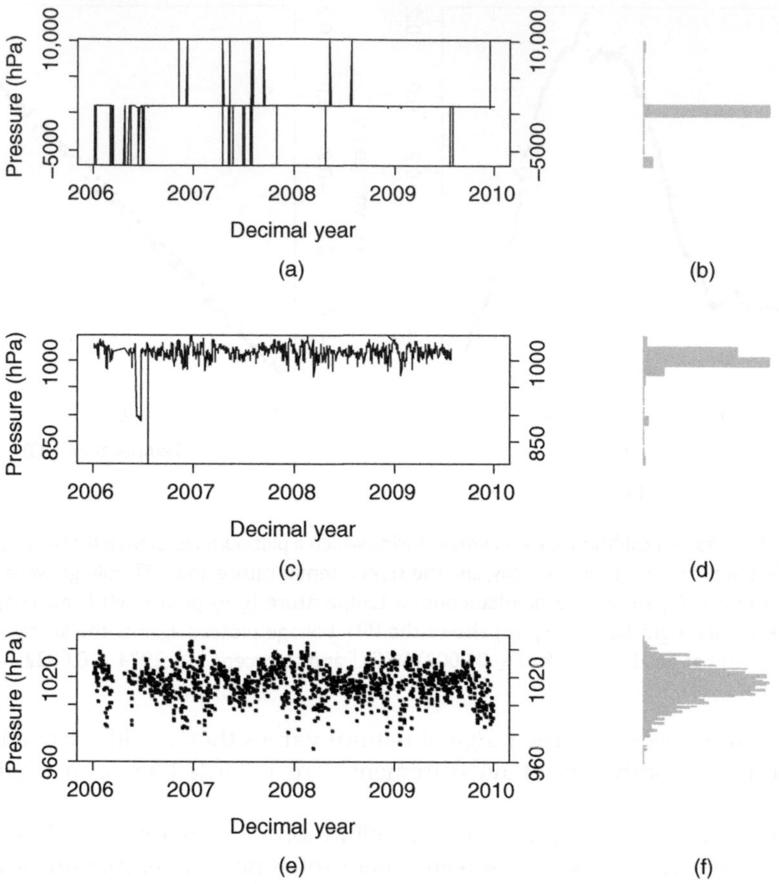

Figure 4.10 Daily time series of mean sea level (msl) pressure (a) as presented from a raw data file (with associated rotated histogram of values (b)); (c) and (d), after the out of range values have been removed; and (e) and (f), after the outliers have been removed by restricting the range to values between 950 hPa and 1050 hPa.

After suppressing the plotting of the NA values, Figure 4.10c is obtained, with the associated histogram Figure 4.10d. Many more of the data values lie within a physically expected range, but there are still a few values during 2006 which lie well away from the majority of values. Because line plotting is used, attention is drawn to the anomalies presented by outliers. If these outliers can be confidently[xiv] regarded as totally unphysical (e.g. lying beyond reliably known extreme values), these can be removed by assigning upper and lower limits on the data. After removing these, Figure 4.10e shows the cleaned-up time series for further analysis. The associated histogram (f) now shows a distribution much more characteristic of physical

[xiv] This requires careful judgement as the variability in the natural world can be completely surprising. The 1984 discovery of the Antarctic ozone hole (when column ozone measurements were exceptional compared with previous measurements), and the deep 2009 solar minimum both indicate caution is wise. The risk always remains of excluding unprecedented events.

Table 4.3 Summary statistics for the pressure data (in hPa) of Figure 4.10

	Minimum	Lower quartile	Median	Mean	Upper quartile	Maximum
Raw data (b)	−6999	1006	1016	621	1023	9999
NA values suppressed (d)	809	1009	1017	1014	1024	1045
Outliers removed (f)	969	1009	1017	1016	1024	1045

measurements, with the majority of values clustered around a central value, with diminishing tails to one or both sides (see also Figure 2.8).

For each of the time series of Figure 4.10a, c and e, representative statistics can be calculated. These are given in Table 4.3. The difference between the median and the mean is particularly apparent. Even with the clearly erroneous values included in Figure 4.10a, the median is almost identical to the median after the removal of outliers Figure 4.10e. The spread of values as given by the range between the quartiles is also hardly affected, although other parameters will show different distributions. In contrast, the mean differs considerably between each of the three cases. Hence a quick test of the raw data values is best carried out by computing the median and IQR, to provide descriptive statistics which are much more robust to outliers than the arithmetic mean.

4.4.3 *Irregular and intermittent time series*

Some time series of data do not contain regularly-spaced values, either because the measurement cannot be obtained due to changes in conditions or opportunities, or because the data logging process has been interrupted. This leads to an irregular time series. As indicated in section 4.3.3, it can be worthwhile constructing a regular, but intermittent, time series from the available values, such as by inserting the NA indication when values are missing. An intermittent time series has some advantages over an irregular time series because the number of entries remains constant, which can simplify subsequent data processing, such as in constructing average values during specific interval. One approach to forming a regular but intermittently populated time series from an irregular time series is to generate a set of evenly spaced time values, alongside a data array of the same size initialised so contain NA values. The original irregular time series is then searched for time values equal (or very close) in time to the evenly spaced time values, and the data array's NA values replaced with real measurements when they are present.

4.4.4 *Further data analysis*

Clearly, further analysis after the data files have been generated, corrected and calibrated will be application-specific, and depend on the scientific questions to be considered. Suggestions for some possible approaches are given in Chapter 12.

5

Temperature

Although many methods exist with which to determine temperatures of liquids and solids, the measurement of air temperature is complicated by the poor thermal conductivity of still air and the presence, during the day, of solar radiation which directly heats thermometers. Consequently, air temperature measurement depends not only on the thermometer employed, but on its exposure. In particular, the uncertainty resulting from solar radiation heating up a thermometer is known as the *radiation error* (see Section 5.5), which causes a thermometer to become warmer than the air to which it is exposed. This effect can cause large uncertainties, and is present in almost all daytime surface and airborne temperature measurements. Thermometer exposure is therefore an important aspect of air temperature measurement in general, requiring a combination of shade, protection from precipitation and good ventilation (see Section 5.5.2). Two methods of thermometry operating on different principles which do not require efficient conduction between the air and the sensor are acoustic thermometry (see Section 8.2.6) and infrared thermometry (see Section 9.7.5).

Temperature difference determines the direction of heat flow between two bodies in thermal contact. There is no heat flow between two bodies if they are in thermal equilibrium, when their temperatures will be equal. Measurement of temperature depends on the existence of thermometric properties of substances, such as the expansion of solids or liquids with temperature or a change in resistance. Many physical properties of solids provide such a thermal response, however, the abundance of different thermometric properties can also complicate a temperature measurement, if other temperature-dependent changes occur in addition to the primary thermometric property sought. An example would be a device using resistance as its primary thermometric property, in which both the sensing element *and* the connecting wires showed a change in resistance with temperature. These and similar errors can usually be reduced or eliminated by good experimental technique.

5.1 The Celsius temperature scale

The standard international temperature scale is based on properties of an ideal gas, leading to the thermodynamic temperature scale. The thermodynamic (Kelvin) scale

Meteorological Measurements and Instrumentation, First Edition. R. Giles Harrison.
© 2015 John Wiley & Sons, Ltd. Published 2015 by John Wiley & Sons, Ltd.
Companion website: www.wiley.com/go/harrison/meteorologicalinstruments

uses the triple point[i] of water as a fixed point, with the divisions of the scale made equal to (1/273.16) of this temperature. Each of these divisions is known as 1 K.

Temperatures measured in Kelvin are needed in evaluating thermodynamic properties of atmospheric gases, but for surface air temperature measurements, the Celsius scale often becomes more practical. The Celsius scale shifts the thermodynamic scale by a fixed offset, so that a temperature T_K (in Kelvin) is given as T_C (in Celsius) by

$$T_C = T_K - 273.15. \tag{5.1}$$

Because there is only an offset between T_C and T_K, the size of the temperature unit is the same in Kelvin and Celsius. Using the Celsius scale, the triple point of water is 0.01°C, the melting point of ice and boiling point of water are about 0°C and 100°C respectively depending on the pressure, and absolute zero (0 K) is –273.15°C.

5.2 Liquid in glass thermometry

5.2.1 *Fixed interval temperature scales*

Early temperature scales used the melting and freezing points of convenient materials to define temperature scales empirically, for a particular thermometer (see also Section 1.4). To determine temperatures lying between these fixed points, the thermometer's scale was divided equally. Such an empirical temperature scale is necessarily specific to the thermometer in use, because of non-linearity in the thermometric property employed. Hence comparison between temperatures measured on two empirical scales using the same fixed points but different thermometric properties is difficult if high precision is required.

An important empirical temperature scale is the Centigrade scale. This employs one hundred equal divisions between two assigned temperatures, assuming that the thermometric property employed is linear. Therefore, if the thermometric property has some value X which lies between the lower fixed point X_0 and upper fixed point X_{100}, the temperature associated with the value X will be

$$\theta(X) = 100\frac{(X - X_0)}{(X_{100} - X_0)}. \tag{5.2}$$

If the lower and upper fixed points are (as is usual) the melting point of ice[ii] and the boiling point of water respectively, at one atmosphere pressure, the Centigrade and Celsius scales will, despite their fundamentally different origins, show approximate numerical agreement if non-linearity in the thermometric properties is small.

---

[i] This is the unique temperature at which the vapour, liquid and solid phases of water co-exist in equilibrium. This temperature is conveniently obtained through the use of a triple point cell, which is a cooled cylindrical glass jacket containing melting ice. A thermometer is inserted and read when solid, liquid and vapour are simultaneously present.

[ii] Because of the possibility that small quantities of pure water can remain liquid well below the freezing point of bulk water (i.e. through supercooling, see Figure 5.6), the melting point of ice is chosen as less ambiguous.

5.2.2 *Liquid-in-glass thermometers*

The fixed interval temperature scale is closely associated with the liquid-in-glass (LiG) thermometer which is still widely used for general-purpose measurements in air or liquids. A LiG thermometer uses thermal expansion of a liquid as its thermometric property, with alcohol or mercury the liquid commonly used in thermometers for atmospheric applications. The cubic thermal expansivity of a liquid α is the fractional change in the liquid's volume for a unit change in temperature. For a volume of liquid V_0 at a temperature T_0, the change in volume ΔV when the temperature has changed to T is given by

$$\Delta V = \alpha V_0 (T - T_0). \tag{5.3}$$

If the liquid is constrained within the glass tube of the thermometer, its expansion can be measured from a scale ruled on the glass (Figure 5.1).

A LiG thermometer typically has a cylindrical geometry, that is, the expansion or contraction of the liquid occurs along a glass bore of constant circular cross section. For a bore of radius r, the sensitivity of the thermometer (the change in length of the liquid thread per unit change in temperature) is

$$\frac{\Delta l}{\Delta T} = \frac{\alpha V_0}{\pi r^2}, \tag{5.4}$$

where V_0 is the volume of liquid contained at a reference temperature T_0. A sensitive thermometer therefore requires thin bore glass tubes and/or a large reservoir of liquid in the sensing bulb. To encompass a wide range of temperatures, the stem may need to be long. In variants of LiG thermometers used for soil temperature

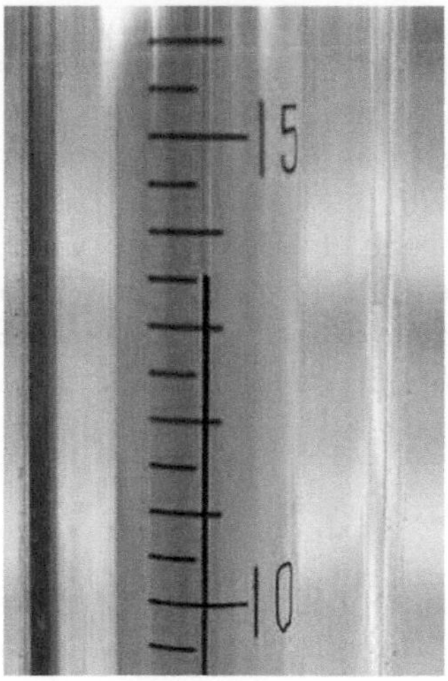

Figure 5.1 The scale engraved on the outer surface of a liquid-in-glass mercury thermometer. This type of thermometer can be read to ± 0.1°C (in this case it is indicating 13.5°C), but whether this is justified depends on its calibration.

Table 5.1 Properties of liquids used in thermometry

Liquid	Cubic expansivity (10^{-3} K^{-1})	Melting point
Alcohol	1.12	−114°C
Mercury	0.81	−39°C
Water	0.21	0°C

Table 5.2 Summary of uncertainties associated with liquid-in-glass thermometers

Aspect	Comment
Scale errors	Arise from a combination of non-linear expansion of the liquid or non-uniformity in the bore; minimised by calibration.
Thread errors	Breaks occurring in liquid thread, which may arise in alcohol thermometers.
Emergent stem error	During calibration, a thermometer may have its bulb immersed in a fluid but with its stem emerging, leading to a temperature gradient along the glass of the thermometer.
Parallax errors	Arise from refraction within the glass; observer's eye level should be at the same level as the liquid level.
Radiation errors	Reduced by using radiation shields, such as polished metal, and by ventilating the sensor (see Section 5.5).

measurements, the sensing bulb is manufactured to be at right angles to the scale, allowing the thermometer to be read without removing it from the soil.

Alcohol and mercury are commonly used for thermometry, and their properties are summarised in Table 5.1. Although alcohol has the larger expansivity, mercury is preferred as it provides a well-defined meniscus, and does not adhere to the glass of the thermometer body. Mercury is also opaque, whereas alcohol has to have a dye added to provide contrast with the glass body of a thermometer. A practical disadvantage of mercury (other than its toxicity) arises because it freezes at −39°C. Alcohol thermometers consequently have to be used for LiG temperature measurements below this temperature, and for minimum thermometers.[iii] These can show poorer long-term stability than mercury thermometers.

Accuracy of LiG thermometers is typically ±0.2°C, even though their resolution may be better, for example, with 0.1°C divisions. The typical exponential response time is 30 s, so about 100 s is needed to give a steady reading (see Section 2.2.1). This time scale depends on the size of the sensing bulb, which, as well as influencing the sensitivity, determines the thermal capacity of the device. Some sources of uncertainty specific to LiG thermometers are given in Table 5.2.

5.3 Electrical temperature sensors

Electrical thermometers of different kinds present a range of alternative technologies to the LiG thermometers, through providing a voltage (e.g. a thermocouple or

[iii] Minimum or maximum thermometers exist based on LiG thermometers, which retain the extreme value of temperature reached until it is reset. In modern variants of the combined maximum and minimum thermometer originally designed by James Six (1731–1793) in 1782, the extreme temperatures reached are marked by a metal index moved by the mercury column, which can be reset using a magnet.

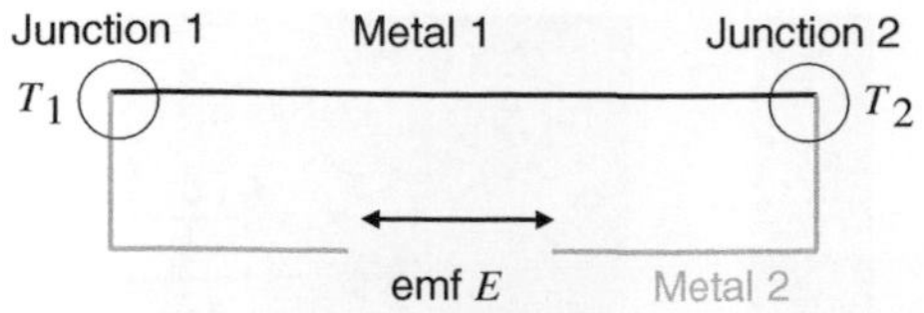

Figure 5.2 A thermocouple, constructed by making junctions 1 and 2 between two dissimilar metals. The emf generated E is proportional to the temperature difference ($T_1 - T_2$) between the two junctions.

semiconductor thermometer) or resistance (e.g. a thermistor or metal resistance thermometer) output which varies with temperature.

5.3.1 *Thermocouple*

A thermocouple junction results when two different metals are welded, twisted or soldered together. If two such thermocouple junctions are electrically connected in series, a small emf is generated[iv] which varies with the temperature difference between the two junctions (Figure 5.2).

The choice of metals (or alloys) used determines the emf generated. Common combinations of metals used for thermocouples include copper–constantan (known as 'type T' thermocouples) chromel–alumel ('type K') and iron–constantan ('type J'), which all give emfs of about 40 $\mu V K^{-1}$. The temperature-voltage characteristic of a thermocouple over a large range is non-linear, in which case it is desirable to operate at the temperature around which it is most sensitive. A thermocouple thermometer can be made physically small, requiring no more than a short section of the dissimilar wire in contact (Figure 5.3).

Thermocouple junctions are usually connected differentially in pairs, with one kept at a fixed reference temperature (usually 0°C), or an electronic reference junction used to provide a fixed voltage. Thermocouples are therefore ideally suited to determining temperature differences, by exposing the two junctions to two different temperatures. The response time depends on mass and ventilation speed and is typically 0.5 s to 5 s. A major limitation is that the small emf produced requires a sensitive voltmeter; for example, a voltmeter reading to 1 μV is required to give a resolution of 0.025°C for a type K thermocouple. Long, or unscreened connection leads can generate noise voltages comparable with the thermocouple voltage, compromising the measurement (see also Section 3.3). Individual thermocouples can, however, be connected in series to form a thermopile, which, as a system gives a larger output and is more sensitive than a single thermocouple, but may be physically larger.

5.3.2 *Semiconductor*

Many modern semiconductor temperature sensors are available, based on the variation of forward voltage across a semiconductor *pn* junction, such as that of a diode. A

[iv] Generating a thermocouple emf is an example of the *Seebeck effect*. If, instead, a current is applied, one junction will cool with respect to the other, which is known as the *Peltier effect*. Qualitatively, both effects are associated with a thermal gradient across the metal junction, causing charge carriers at the hot end of a metal to have more energy than those at the cold end, with the excess energy exchanged with the metal lattice.

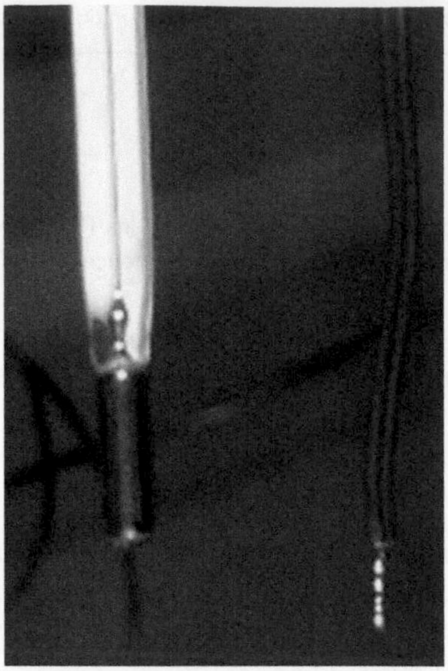

Figure 5.3 A simple thermocouple junction, constructed from two dissimilar wires twisted together.

simple silicon diode junction has a change in forward voltage (Figure 5.4) with temperature of $\sim -2\ \mathrm{mV\,K^{-1}}$, which forms the basis for many integrated semiconductor temperature sensors.

As amplification circuitry can also be fabricated on the same semiconductor chip, amplification to give convenient linear sensitivities can be implemented, such as to generate a change of $10\ \mathrm{mV\,K^{-1}}$ for easy use with a digital voltmeter. Some semiconductor temperature devices also include analogue to digital converters, to provide a digital output directly. Using serial data transfer (one data bit sent after another, at a prearranged rate), it is possible for the sensor to need no more than three connections, two for power and one to provide the data.

Accuracy of semiconductor temperature sensors can be to about ±1°C or better without further calibration. Because of their insulating packaging material, semiconductor temperature sensors have response times which are somewhat longer than those of thermistors.

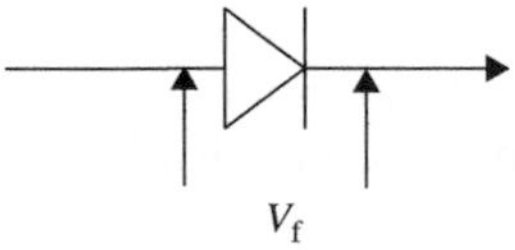

Figure 5.4 Electronic circuit symbol for a diode (anode connection on left, cathode connection on right), showing the forward voltage (V_f), which is developed across the device when a current flows through it in the direction of the arrow head.

5.3.3 *Thermistor*

A thermistor is a semiconductor device specifically fabricated so that its electrical resistance varies markedly with temperature. Thermistors may have a positive or, more commonly, a negative temperature coefficient.[v] A thermistor with a negative temperature coefficient can have its resistance characteristic represented by

$$R(T) = a \exp\left(\frac{b}{T}\right), \tag{5.5}$$

where R is the resistance of the device at a temperature T in Kelvin, and a and b are constants for a given material. (b is sometimes known as the characteristic temperature). The exponential dependence of resistance on temperature means that, over some ranges of temperature, a thermistor can be very sensitive i.e. its resistance changes rapidly per unit change in temperature. Thermistor resistances are usually sufficiently large to be easily measured on a simple ohmmeter, but the scale response will be non-linear.

Thermistors require standardisation to determine the constants a and b, but if the thermistors are manufactured to sufficiently close tolerances, their constants can be assumed from data sheets without a separate calibration experiment. In common with other sensors, they may need ventilation and radiation shields for use in air, but they can be made physically small. The response time depends on their mass and the ventilation speed used. It is typically 1 s to 10 s, with typical resolutions of 0.02°C. For long-term use, a thermistor has to be aged: this process ensures that any initial changes in a thermistor occur before it is put into service.

An example of the use of an ultra-miniature thermistor for temperature measurement within a water droplet is shown in Figure 5.5. The thermistor provides the mounting for the droplet, whilst measuring its temperature during supercooling. (In the original experiment [42], a heating current was applied to the thermistor after freezing had occurred, to initiate melting.) Figure 5.6 shows the variation of the droplet temperature during a cooling cycle.

5.3.4 *Metal resistance thermometry*

Resistance thermometry depends on the property shown by metals of an increase in resistance with temperature. As platinum has a well-defined (and substantially linear) relationship between resistance and temperature and is chemically stable, it is well suited to thermometry. The metal resistance sensing element of a Platinum Resistance Thermometer (PRT) can be engineered into a form suitable for the intended application: high-quality resistance thermometers use a coil of platinum wire (which is stable and resistant to corrosion) in a metal case, and cheaper versions use a platinum film deposited on a ceramic substrate.

The general form of the temperature-resistance characteristic[vi] is

$$R(T) = R_0 \left[1 + \alpha(T - T_0) + \beta(T - T_0)^2\right], \tag{5.6}$$

where R_0 is the resistance at a temperature T_0, and α and β are measured temperature coefficients for the metal concerned. For accurate work, the quadratic term $\beta(T - T_0)^2$

[v] A negative temperature coefficient device has a response in its thermometric property (resistance in the case of a thermistor) which decreases with increasing temperature.

[vi] This is a simplified form of the Callendar–Van Dusen equation for platinum resistance thermometry, which arose from the work of the Gloucestershire-born physicist Hugh Longbourne Callendar (1863–1930). His son, Guy Stewart Callendar (1898–1964) pioneered the idea that rising atmospheric carbon dioxide concentrations influence the global temperature.

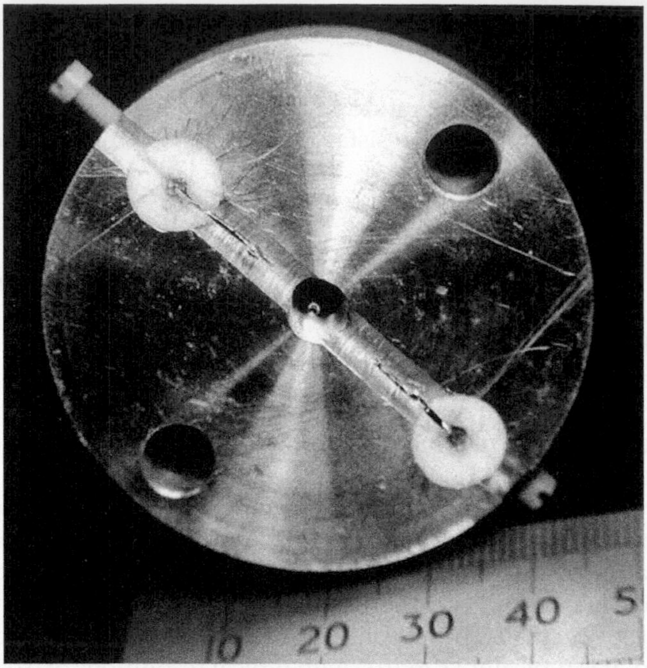

Figure 5.5 Droplet cooling apparatus using an ultra-miniature glass bead thermistor to both support and monitor a water droplet. The complete aluminium cooling chamber has been dismantled to expose the position and mounting of the water droplet on the upper aluminium block of a cooling chamber. A droplet is in the centre of the picture, held on the bead thermistor strung on its fine axial connection wires between the thick connection wires emerging from PTFE slugs. (In use, the thermistor was covered by a lower aluminium block, spaced from the thermistor and cooled by a Peltier device.)

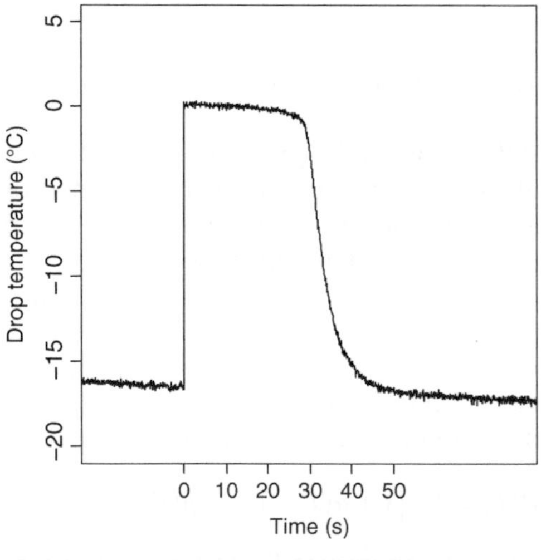

Figure 5.6 Temperature within a 1 μL water droplet, as the surrounding environment was steadily cooled. The droplet remains liquid (i.e. it supercools) until spontaneously freezing (at time = 0 s), during which its temperature instantaneously rises to about 0°C as latent heat is released. After the droplet has completely frozen, it cools to the temperature of its surroundings, at a rate proportional to the difference between its temperature and that of its surroundings. (Temperature measurements made at 12 Hz.)

Figure 5.7 Platinum resistance thermometer (resistance element enclosed within a stainless steel sheath, 8 mm diameter), used to measure the temperature of a concrete surface. (In use, a half-cylinder radiation shield is fitted into the clips.)

is required, but in many cases, using platinum this can be neglected as the term is small ($\beta \sim -5 \times 10^{-7}\ \mathrm{K}^{-2}$). Widely-used commercial PRTs conform to the industrial 'Pt100' standard, having a resistance of 100.0 Ω at 0°C with a temperature coefficient α of 0.00385 K^{-1}, and good long term stability without further standardisation. Their response time depends on the size and mass of the case, but is typically 30 s to 120 s. The typical temperature resolution is 0.02°C, with 0.05–0.2°C accuracy possible if used carefully. A commercial PRT sensor is shown in Figure 5.7.

For air temperature measurements, a cylindrical PRT sensor would usually be used with a radiation shield of some form. However, because platinum can be made into fine wire, it is also possible instead to construct fine wire PRTs which present a much reduced area for the interception of solar radiation and small thermal inertia allowing rapid time response. Such a fine wire sensor [43] was used on the Huygens probe which landed on Saturn's moon Titan in 2005. In versions for terrestrial micrometeorology, there are similar objectives of small radiation errors and rapid time response. With fine wire sensors, it is also important to limit the signal conditioning circuitry's measurement current which passes through the fine wire sensor, to minimise errors from self-heating.

Figure 5.8 shows a fine wire platinum resistance thermometer (FWPRT) element, constructed from platinum wire wound on a plastic former. Connections for the resistance measurement are made to each end of the fine wire by a soldered joint, and additional connections are made to allow for compensation of the connection resistances. Figure 5.9 shows the same design of FWPRT used in atmospheric conditions, mounted in an array for comparison between the sensors. Rapid fluctuations in temperature can be obtained with such sensors, as the exponential time response is ~40 ms. The time response capability of a fine wire device for rapid changes in temperature is apparent in the measurements presented in Figure 5.10. Such thermometers are useful for studying turbulence, and for measuring the transfer of heat by turbulence (see Section 12.1.2).

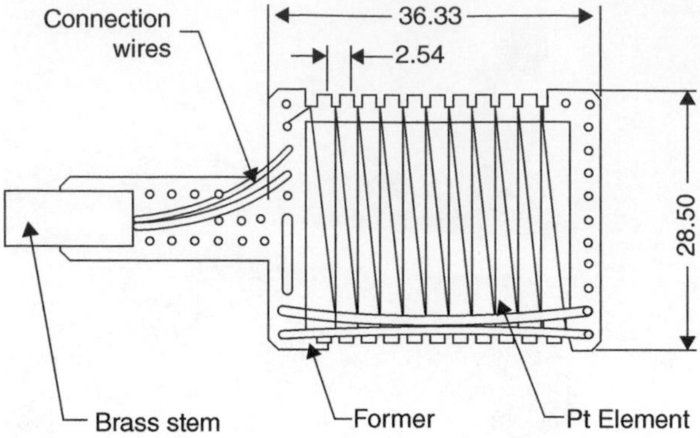

Figure 5.8 Construction of a fine wire platinum resistance thermometer, using fine (25 μm diameter) platinum wire [37] (dimensions given in millimetres).

5.4 Resistance thermometry considerations

For continuous logging of temperatures, a voltage rather than a resistance measurement will be needed. Resistance to voltage conversion is, in principle, a straightforward process but, as mentioned in Section 3.5, the accuracy of a resistance measurement depends on the magnitude of the resistance to be measured in comparison with

Figure 5.9 Array of fine wire resistance thermometers arranged for an air temperature comparison. The signal conditioning electronics is mounted in the box at the base of the thermometer. This minimises the effect of connection resistances, but still requires good thermal stability of the electronics.

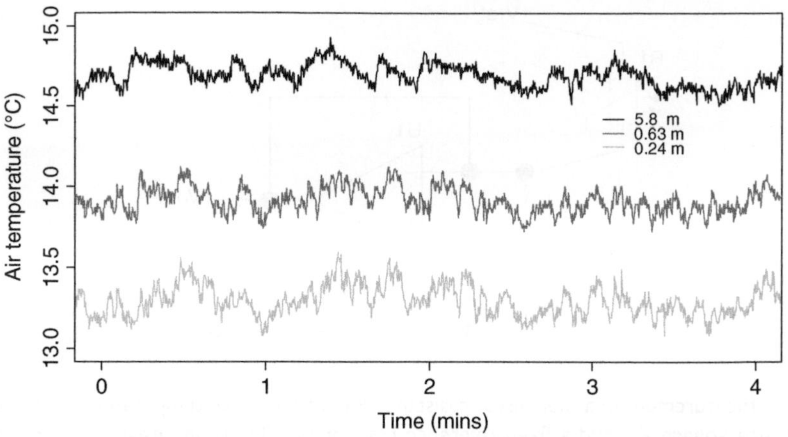

Figure 5.10 Air temperature measurements obtained over flat terrain (Camborne Met Office, Kehelland, 11 August 1999), using fine wire platinum resistance thermometers in a vertical array at 0.24 m, 0.63 m and 5.88 m above the surface.

connection resistances. The linearity of the sensor can also be a consideration, if the measurement resolution is required to be consistent across the entire measurement range.

5.4.1 *Thermistor measurement*

A key practical advantage of thermistors is that their resistances are large compared with the connection cables required, which means that they are well suited to temperature monitoring remote from the signal conditioning circuitry. However, because of their exponential response, some linearisation is desirable. The non-linear response to resistance seen in the Figure 3.19 resistance measuring circuit can be used to approximately compensate for the thermistor's exponential response. For a thermistor with a negative temperature coefficient (ntc), connecting the thermistor in the upper part of the potential divider with the resistor in the lower part (Figure 5.11) will cause the output voltage to increase with increasing temperature.

The full theoretical voltage-temperature response can be found by combining the thermistor characteristic (Equation 5.5) with that for the potential divider circuit (Equation 3.10), as

$$v_o = V_{\text{ref}} \frac{R}{a \exp\left(\frac{b}{T}\right) + R}, \tag{5.7}$$

where R is the fixed series resistance (R2 in Figure 5.11), V_{ref} the reference voltage and a and b the thermistor coefficients. The change in the output resistance with temperature for a typical ntc thermistor can be seen in Figure 5.12a. From

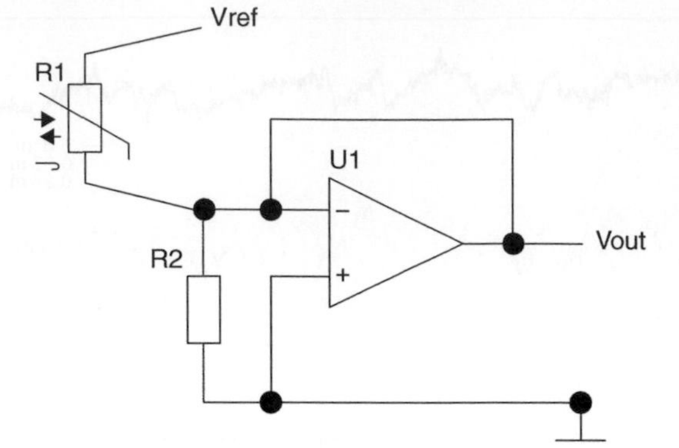

Figure 5.11 Measurement of a thermistor resistance R1 with negative temperature coefficient, using a fixed reference voltage V_{ref} and a fixed reference resistor R2. (U1 is an integrated circuit unit gain buffer stage.)

Equation 5.7, the variation in output voltage v_0 with temperature is given by

$$\frac{dv_o}{dT} = V_{\text{ref}} \frac{Rab \exp\left(\frac{b}{T}\right)}{T^2 \left[a \exp\left(\frac{b}{T}\right) + R\right]^2}. \tag{5.8}$$

By optimising the choice of R, the variation with temperature of the temperature response dv_0/dT can be reduced [44]. For R about equal to the thermistor resistance in the middle of the temperature measurement range sought, the voltage-temperature

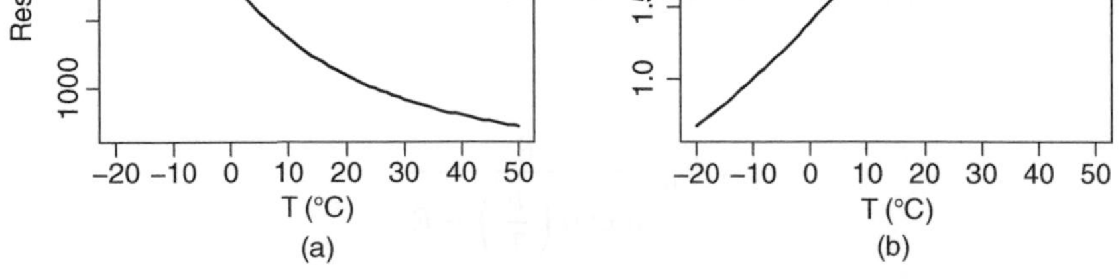

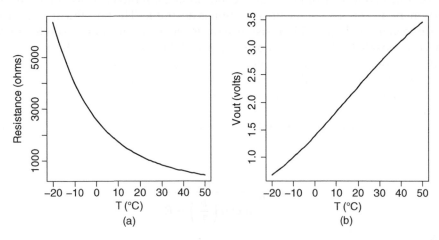

Figure 5.12 (a) Calculated resistance variation with temperature T for a negative temperature coefficient thermistor having a resistance of 1 kΩ at 298 K. (b) Calculated output voltage variation with temperature for the thermistor in (a) in the non-linear resistance measuring circuit of Figure 5.11, for R2 = 1 kΩ and V_{ref} = 5.0 V.

characteristic of the circuit differs from a linear response by less than 1°C between −15° and 40°C (see Figure 5.12b).

5.4.2 *Platinum resistance measurement*

In contrast to typical thermistors, the resistance change with temperature for Pt100 sensors is highly linear through the atmospheric range of temperatures encountered, but only small (e.g. a 0.385 Ω change in resistance for each 1 K temperature change). Care must therefore be taken to allow for finite connection resistances if the sensor is remote from the signal conditioning electronics. An additional consideration is self-heating of the sensor from the measurement current, which must be minimised.

Figure 5.13 shows a practical resistance-to-voltage converter [45], using the Kelvin connection method (see Section 3.5.2) for a Pt100-like sensor whose resistance varies around a base resistance of 100 Ω at 0°C. An accurate 5 V voltage source supplies a fixed excitation current of about 50 μA through R1, R2 and the sensor resistance R, although, as (R2+R) ≪ R1, R and R2 contribute little to determining the excitation current. The excitation current develops a fixed voltage across R_2, which is amplified by a factor of 100 by a precision microvolt differential stage (IC4). A further microvolt differential stage (IC5) amplifies the voltage generated across the sensor resistance, which will vary with temperature, by the same amount. The difference between the outputs of IC4 and IC5 is determined by differential amplifier IC6a, which will vary

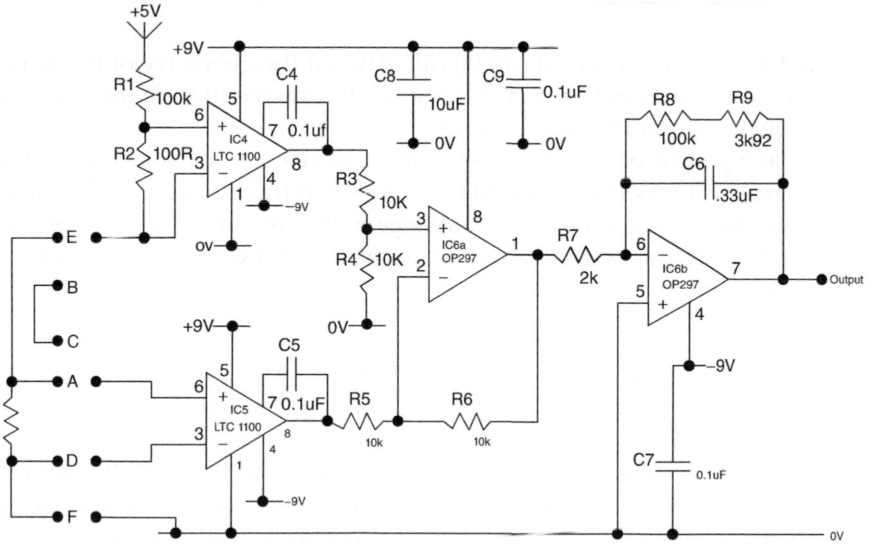

Figure 5.13 Measurement circuit for a platinum resistance thermometer, which allows for connection resistances to a sensing element of resistance *R*. The resistance element provides four connections A, E, D and F, with one pair (E and F) providing the excitation current, and the other pair (A and D), access for a voltage measurement (B and C provide a shorting link for use with alternative resistance compensation circuitry.). R1 and R2 are precision resistors, chosen for temperature stability and close tolerance.

with R. Further amplification is provided by IC6b to ensure the output goes positive with increasing resistance, with the gain chosen to give an output voltage change suitable for the resistance change envisaged.

5.5 Thermometer exposure

Solar radiation falling on a thermometer used for air temperature measurements will cause its temperature to be greater than true air temperature, causing a *radiation error*. Even for the fine wire sensors of Figure 5.10, a frame and body of non-negligible area are still required to support the fine wire element. The radiation error presents a fundamental problem for accurate air temperature measurements.

5.5.1 Radiation error of air temperature sensors

For a thermometer in thermal equilibrium with its surroundings (when its temperature is stable), the radiant energy received by the thermometer balances that lost by thermal radiation and convection, that is, when

Rate of acquisition of energy	=	Heat loss rate by radiation	+	Heat loss rate by convection

In principle, the energy received can be calculated if the geometry of the sensor is known, as this determines both the effective areas for interception of the solar radiation and for convective and radiative losses.

Consider a temperature sensor exposing an area A to radiation, constructed of material with reflection coefficient α and emissivity ε. If there is no conduction away from the device, then, for an incident irradiance S, the rate of acquisition of energy is $S\,(1-\alpha)\,A$. Balancing this source term with terms for the heat loss rate by radiation and convection gives

$$S(1-\alpha)A = \varepsilon A'\sigma(T^4 - T_a^4) + \frac{k(T - T_a)}{d}A'N, \tag{5.9}$$

where the first RHS term represents radiative loss (from the Stefan-Boltzmann Law, with σ the Stefan–Boltzmann constant) and the second, the convective loss. Both the radiative and convective losses are assumed to occur from the same area A', which, in general, will differ from the area able to intercept the radiation A. The convective loss is found from the difference in temperature between the sensor temperature T and air temperature T_a and thermal conductivity of air, for which the constant of proportionality – the convective heat transfer coefficient – is known as the Nusselt number. The Nusselt number N is given by $N = c(\mathrm{Re})^{\zeta}$, where c and ζ depend on geometry, and Re is the Reynolds number. Re is calculated as Ud/v where U is the

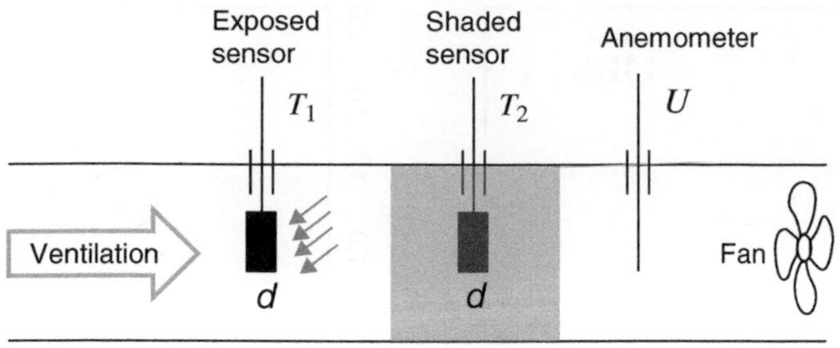

Figure 5.14 Experiment to compare the effect of radiation on two identical cylindrical sensors of diameter *d*, one (T_1) exposed to radiation, and the other (T_2) shaded, in ventilation of variable speed *U*.

air speed past the sensor, *d* the sensor diameter and ν is the kinematic viscosity of air. For a cylinder, $c = 0.62$ and $\zeta = 0.5$. Substituting for the Nusselt number gives

$$S(1-\alpha)A = \varepsilon A'\sigma(T^4 - T_a^4) + \frac{k(T-T_a)}{d}A'c\sqrt{\frac{Ud}{\nu}}. \quad (5.10)$$

Of the two loss terms on the right-hand side, the heat loss by thermal radiation is much smaller, and can be neglected in comparison. Hence the difference in temperature between the air and the sensor – this is the radiation error ΔT – is, for $A = A'$,

$$\Delta T = (T - T_a) = \frac{S(1-\alpha)d}{kN} = S\frac{(1-\alpha)\sqrt{\nu}}{kc}\sqrt{\frac{d}{U}}, \quad (5.11)$$

or,

$$\Delta T \propto S\sqrt{\frac{d}{U}}. \quad (5.12)$$

This shows that the radiation error depends directly on the amount of non-reflected radiation, but is also proportional to the sensor diameter (as $d^{1/2}$ for a cylinder), and inversely proportional to the wind speed (as $U^{1/2}$ for a cylinder). Laboratory experiments (Figure 5.14) using a pair of cylindrical sensors exposed to a radiation from a bright lamp under varying ventilation (Figure 5.15), confirm the form of Equation 5.12.

These laboratory experiments show that, without ventilation, sensors with appreciable areas for interception of solar radiation in the atmosphere may readily record temperatures which are several degrees greater than the local air temperature.

5.5.2 *Thermometer radiation screens*

To minimise radiation errors in the measurement of air temperature for long-term climatology purposes, it has long been recognised (see Section 1.4) that an instrument

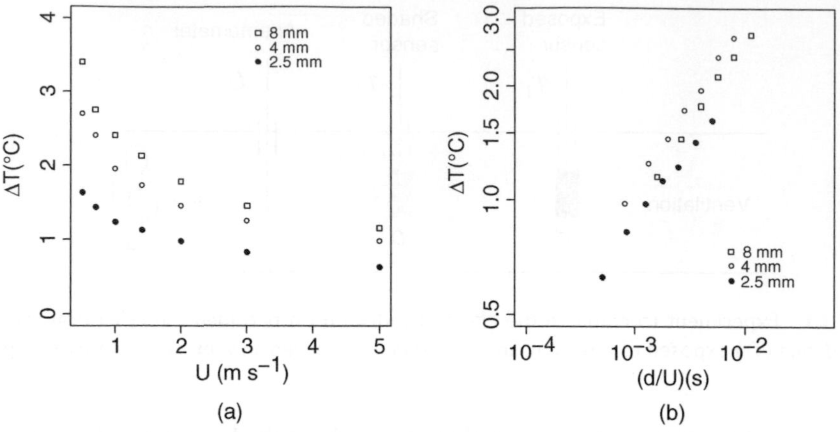

Figure 5.15 (a) Difference in temperature (ΔT) between a pair of identical blackened cylinders, one shaded and the other exposed to radiation from a 700 W tungsten lamp, with the air speed U across the cylinders varied. The experiment was repeated for pairs of sensors having diameters 2.5 mm, 4 mm, and 8 mm. (b) data from (a) plotted on a log–log plot, as (d/U) versus ΔT. (For the 8 mm diameter sensor, $\zeta = 0.46$, using a least-squares fit.)

shelter or radiation screen of some form is needed to protect the thermometer's sensing element from direct solar radiation. The basic requirements for an ideal thermometer shelter [46] are that it should:

1. provide shielding from direct rays of the sun at all times;
2. not affect the thermometers by warming up;
3. prevent reflected radiation from reaching the thermometers;
4. exclude external sources of heat (e.g. from buildings);
5. allow free passage of air around the thermometers.

Instrument shelters can of course also offer protection from precipitation for thermometers and other sensors, such as those required typically for humidity measurements.

Some hand-held instruments are specially designed to minimise radiation errors in temperature measurements, by forcing ventilation through a highly reflective housing (e.g. the Assmann psychrometer, Section 6.3.7), or by minimising the radiation interception area, for example through the use of a fine wire thermometer.

The double-louvered wooden screen of Thomas Stevenson is a well-established design of thermometer shelter, with a long and widespread legacy because few changes have been made to the original design across a century. Stevenson screens are still widely deployed at standardised measurement sites (Figure 5.16). The question of the absolute accuracy of the air temperature measurements obtained using such a thermometer screen is difficult, as it clearly requires a reference measurement of air temperature for comparison, and there are competing uncertainties in the comparison experiments required. The lag and radiation errors are generally the most important.

Figure 5.16 Stevenson screens in use at Eskdalemuir Observatory, Dumfriesshire.

5.5.3 *Radiation errors on screen temperatures*

In poorly ventilated conditions, radiation errors can limit the accuracy of air temperature measurements using radiation screens [47], notably in calm and/or sunny conditions. This means that the screen temperature T_{scrn} and the air temperature T_{air} may differ and hence that they cannot always be assumed to be equivalent. An extensive investigation [48] of the properties of a Stevenson screen was undertaken between October 1969 and December 1972 at Kew Observatory, London. This work used an aspirated resistance thermometer as the reference measurement most accurately approximating the true air temperature, 10 m away from the screen. This aspirated thermometer was still subject to small effects from surface reflection of solar radiation, but this was estimated to only contribute a maximum error of +0.1°C to the determination of the air temperature.

In daytime conditions, the differences between the screen temperature (T_{scrn}) and aspirated reference thermometer (T_{asp}) in the Kew experiments exceeded 1°C in winter and 2°C in summer; the mean differences obtained are summarised in Table 5.3. The largest differences found were in September when there were frequent light winds, with the worst case error in the most unfavourable conditions estimated to be 2.5°C. Occasional negative temperature differences were originally thought to be due to evaporative cooling, but full calculations showed this was insufficient to explain the values observed. The absence of an evaporative cooling effect has subsequently been corroborated [49].

During nocturnal measurements, it was found that the screen thermometer temperatures exceeded the aspirated thermometer temperatures (T_{asp}), particularly at low wind speeds (Figure 5.17). In this study, these effects were attributed to the slower response of the Stevenson screen (estimated as ~20 min) compared with the aspirated thermometer (response time ~2 min), in adapting to decreasing temperatures.

Table 5.3 Differences between Stevenson screen and aspirated thermometer reference temperature ($T_{scrn} - T_{asp}$) for the maximum and minimum temperatures [48], by month

Month	$T_{scrn} - T_{asp}$ for maximum temperature (°C)	$T_{scrn} - T_{asp}$ for maximum temperature (°C)
January	0.04	0.11
February	0.16	0.18
March	0.17	0.14
April	0.14	0.15
May	0.28	0.25
June	0.24	0.19
July	0.34	0.21
August	0.28	0.25
September	0.41	0.32
October	0.21	0.15
November	0.01	0.08
December	0.03	0.08

Further insight into the behaviour of a thermometer screen, particularly under low wind speed conditions, can be obtained from comparison of the continuous variation in screen temperature with a more rapidly responding air temperature reference measurement than that available for the Kew study. A fine wire thermometer provides a useful rapid response reference thermometer in open air, but, for accurate air temperature measurements, it must also be screened from direct solar radiation. This can be achieved using a mechanical solar tracker, a device designed to track the sun for solar radiation measurements (see Section 9.4), to keep the thermometer within the shade. Figure 5.18 shows the experimental arrangement used [50]. The fine wire

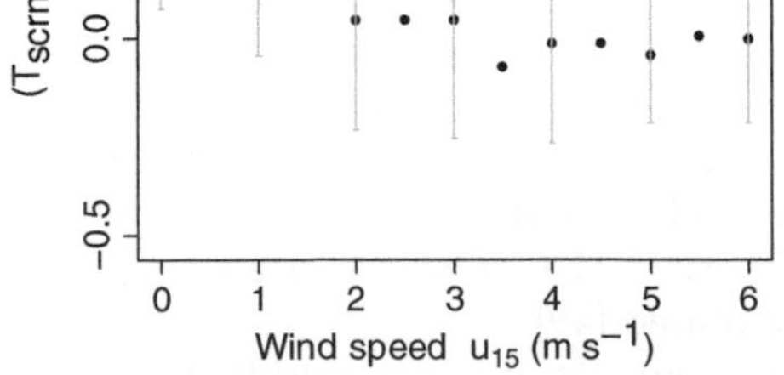

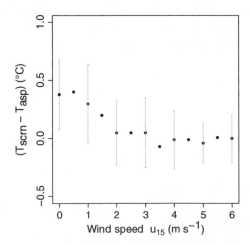

Figure 5.17 Mean differences [48] in daily minimum temperature between the temperature measured within a large thermometer screen (T_{scrn}) temperature and an aspirated thermometer's temperature (T_{asp}), plotted against wind speed at 15 m. (Standard deviations are shown using error bars, when available.)

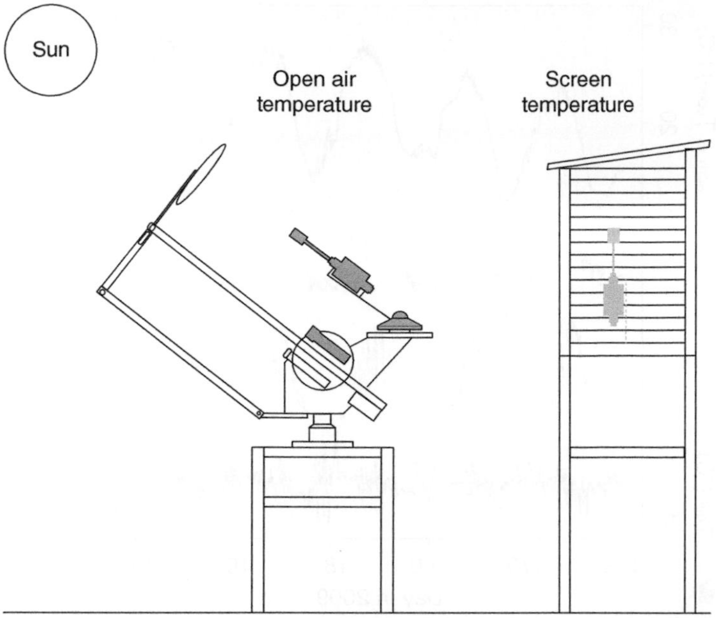

Figure 5.18 Experimental arrangement at Reading Observatory for comparing temperatures determined by a fine wire thermometer in the open air with those determined within a thermometer screen. The open air thermometer (left) is mounted on a mechanical solar tracker, which moves an occultation disk so as to keep the thermometer continuously shaded.

thermometer was regularly compared with the thermometer with the screen to correct for any drift during its exposure.

Measurements of the open air temperature (T_{open}) and the screen temperature (T_{scrn}) obtained during these experiments are shown in Figure 5.19, during fluctuating cloud conditions. The measurements include the solar radiation as determined on a horizontal surface, and the wind speed at 2 m above the surface, near to the thermometer screen. The difference $T_{scrn} - T_{open}$ is also shown. It is clear that this difference is variable, but that it is at its greatest under calm or low wind speed conditions as found in the Kew study. Instantaneously, the difference between the differently-exposed sensors can be over ±1°C, but the median difference is <0.2°C.

On the basis of the Kew study and others, the World Meteorological Organisation [51] indicates the worst case difference between true air temperature T_{air} and screen temperature T_{scrn} in calm and clear conditions can lie between −0.5°C and 2.5°C.

5.5.4 *Lag times in screen temperatures*

Because of the enclosure around them, the response time of the combined screen and thermometer system is fairly slow. The lag time τ (minutes) for a thermometer screen can be represented by an expression of the form

$$\tau = \frac{A}{U^n}, \tag{5.13}$$

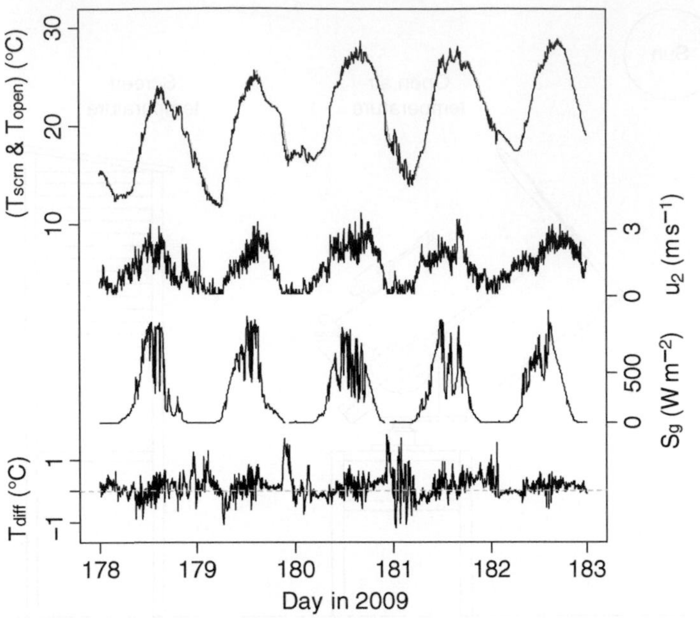

Figure 5.19 Investigation of thermometer screen response in cloudy conditions. Time series (plotted as day of year) showing (upper trace) screen temperature (T_{scrn}, thick grey line) and fine wire thermometer open air temperature (T_{open}, thin line), (second trace) 2 metre wind speed (u_2), (third trace) global solar irradiance (S_g), and (lower trace) screen-air temperature difference ($T_{diff} = T_{scrn} - T_{open}$).

where U is in m s^{-1}. $A = 8.2$ and $n = {}^1\!/_2$ have been reported for a Stevenson screen [52] in wind tunnel experiments, suggesting τ is typically between 5 and 30 min. The screen's lag can also be found from a comparison of the times at which the daily maxima and minima occur in screen temperature with a separate measurement of temperature which is unscreened. Figure 5.20a shows the differences in lag time obtained using this approach for a large thermometer screen, plotted against the wind speed at 2 m obtained nearby using the experimental arrangement of Figure 5.18. The lag time increases substantially at low wind speeds ($u_2 < 1$ m s^{-1}), and the results are reasonably well represented by a power law fit of the form of Equation 5.13. In terms of the temperature variations, at low wind speeds, the lag time leads to a very slight reduction of the amplitude of the daily temperature cycle (Figure 5.20b). This is expected from Equation 2.7, hence the reduction of the amplitude would be greater for more rapid temperature fluctuations, such as that arising from temperature changes generated by the passage of fronts.

To compare the temperature changes which can occur with the screen response, a sequence of air temperatures measured using cylindrical platinum resistance thermometer sensors is shown in Figure 5.21. This demonstrates a fairly rapid change in air temperature, with a temperature change of over 4°C in 30 min, together with the surface wind speed providing the ventilation of the screen. In this case, comparison with Figure 5.20 shows that lag time is much shorter than the timescale of the temperature change.

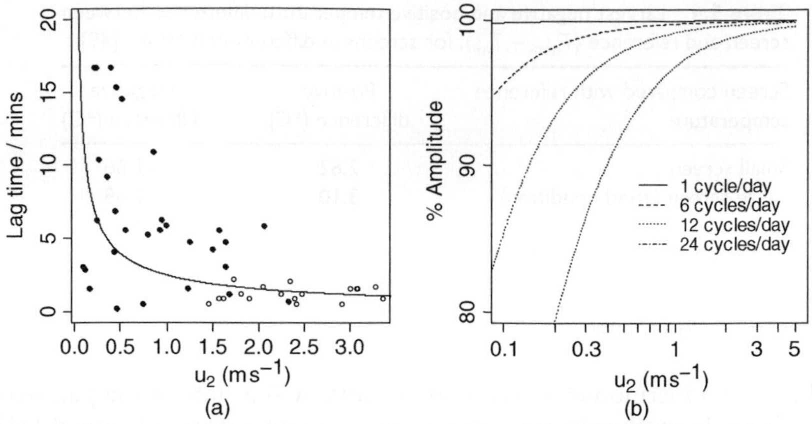

Figure 5.20 Results of a screen thermometer lag experiment [53] at Reading, 2009–2010. (a) Lag time between screen temperatures and an unscreened fine wire temperature, for minimum (solid points) and maximum (hollow points) temperatures, as a function of wind speed u_2, averaged for the hour centred on the screen temperature maximum or minimum. A power law ($\tau = 2.5\ u^{-0.7}$, for u in m s^{-1} and τ in minutes) has been fitted by regression. (b) First order amplitude reduction with wind speed for a range of sinusoidal temperature variations, using the same power law.

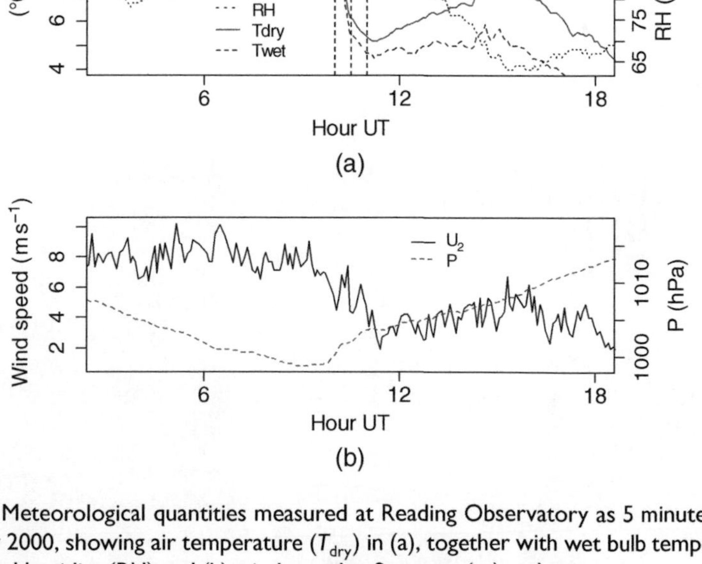

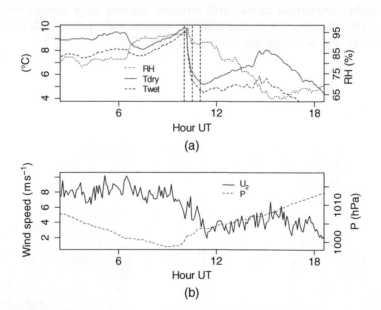

Figure 5.21 Meteorological quantities measured at Reading Observatory as 5 minute mean values, on 10 February 2000, showing air temperature (T_{dry}) in (a), together with wet bulb temperature (T_{wet}), derived Relative Humidity (RH) and (b) wind speed at 2 metres (u_2) and station pressure (P). In (a), the vertical dashed lines mark 1000, 1030 and 1100 UT.

Table 5.4 Largest negative and positive temperature differences between screen and reference ($T_{scrn} - T_{ref}$), for screens in different conditions [49]

Screen compared with reference temperature	Positive difference (°C)	Negative difference (°C)
Small screen	2.62	−1.66
Large screen (good condition)	3.10	−1.68
Large screen (poor condition)	3.61	−2.06

5.5.5 *Screen condition*

The condition of a thermometer screen, most notably the state of the paintwork for wooden screens, is another factor influencing the temperatures observed [54]. This has been studied by comparing one small and two large screens, with an aspirated screen for the reference temperature [49]. One of the large screens was in perfect condition, and the other poorly maintained. The study lasted 207 days, sampling the sensors 12 times per minute to form 1-minute averages (Table 5.4). The screens were all at 1.5 m above the surface within the same enclosure at Norrkoping, Sweden.

The poorly-maintained screen produced the largest deviations, with the small screen having smallest positive and negative extremes, probably due to its more rapid response time and smaller heat capacity. The condition of the screen is therefore another important factor, and wooden screens vary considerably in the maintenance they receive, for example in the state of their paintwork of the upper reflective surface and south-facing louvres (Figure 5.22).

Figure 5.22 Thermometer screen used for climatological observations at a seaside resort.

Figure 5.23 Beehive thermometer screen in use at the meteorological site of the Universitat de les Illes Balears, Palma.

5.5.6 *Modern developments in screens*

For automatic temperature recording systems using data loggers, a wide range of smaller radiation screens of different designs has emerged, generally based on the multi-plate or beehive screen (Figure 5.23). Ventilation is recommended at ~3 m s^{-1}, as long as moisture is not introduced [51]. Plastic Stevenson-style thermometer screens have also been produced, which require greatly reduced levels of maintenance compared with traditional wooden screens. Temperatures obtained in a plastic screen agree to better than 0.25°C with the temperatures obtained in an adjacent wooden screen: the agreement is further improved if the interior of the plastic screen is blackened [55]. Plate or beehive screens are also suitable for locations with limited space, such as on ocean buoys, on which only small thermometer screens can be fitted. Uncorrected, radiation effects on buoy temperature measurements [56] have shown a mean daytime temperature error of 0.27°C, up to a maximum error of 3.4°C.

5.6 Surface and below-surface temperature measurements

5.6.1 *Surface temperatures*

Surface temperature measurements can be obtained from liquid-in-glass or platinum resistance sensors lying on grass, soil, tarmac, concrete or snow surfaces, as needed (see also Figure 5.7). The range of temperatures encountered for such a surface directly heated by solar radiation is, in general, much greater than the corresponding air temperature. Figure 5.24 compares air, soil, grass and concrete temperatures,

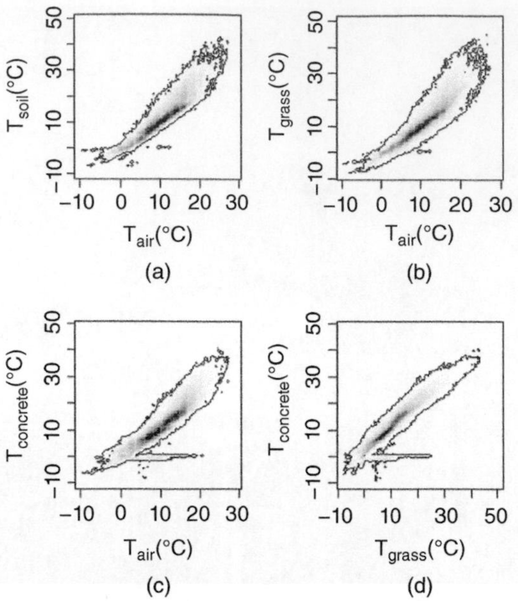

Figure 5.24 Temperatures of (a) soil, (b) grass and (c) concrete surfaces, against air temperature, measured at Reading Observatory between 2007 and 2012. (d) shows the relationship between the grass and concrete surface temperatures. (Plots show increasing density of values in increasingly darker grey tones; the outline contour shows the extreme values.)

showing that the greatest surface temperatures exceed the air temperatures by at least 10°C.

5.6.2 *Soil temperatures*

Beneath the surface, soil temperature measurements at different depths can be obtained from a vertical array of temperature sensors. Figure 5.25 shows measurements made on a cloudless day using a set of thermistors fitted in a soil temperature probe pushed into the soil. Each thermistor was connected to resistance measuring apparatus using a long multi-core cable. The diurnal variation in temperature at the surface becomes suppressed with increasing depth in the soil, and the maximum temperature value occurs later. (The soil temperature's relationship to the vertical heat flux in the soil is discussed further in Section 12.1.3.)

5.6.3 *Ground heat flux density*

The rate of vertical transfer of heat – known as the ground heat flux density – is usually determined using a temperature measurement approach, by finding the temperature difference across a material of known thermal conductivity. Such a device is a *heat flux plate* (Figure 5.26). These are circular or rectangular devices, typically with dimensions of a few centimetres.

A heat flux plate consists of two aluminium plates, sandwiching a resin of known thermal conductivity. By using a thermocouple to determine the temperature

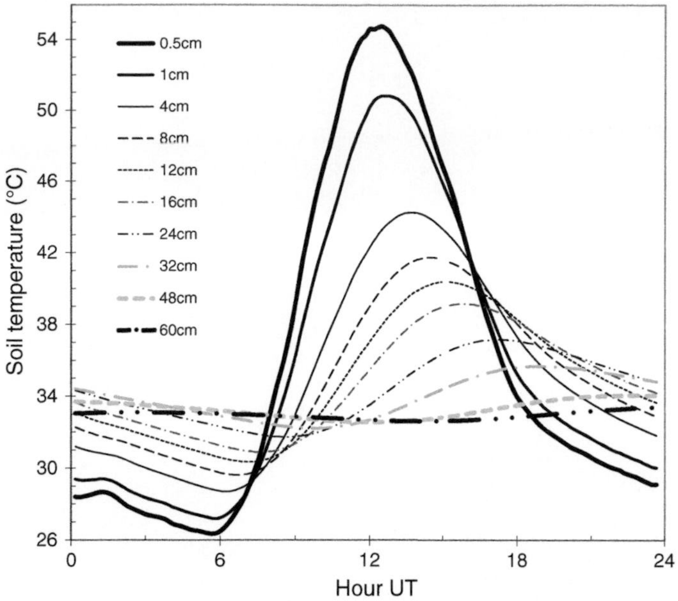

Figure 5.25 Diurnal variation of soil temperatures with depth, obtained during a cloudless day in the Sahel region of Africa (25 September 1992). The temperatures were measured as 10-minute average values, using a set of calibrated bead thermistors, vertically spaced on a wooden probe driven into the soil with the thermistor resistances measured remotely using a multi-core cable and a data logger (see also Figure 9.29).

difference ($T_1 - T_2$) between the upper and lower surfaces, the heat flux can be determined by Fourier's Law. The thermocouple output voltage is sufficiently small that it will often require amplification. The thermocouple voltage is found to be directly proportional to G, and can be calibrated by applying known heat fluxes in thermal equilibrium. Uncertainties in the data from ground heat flux plates arise because their presence disturbs the heat and gas flow in a soil. There may also be poor thermal contact, and differences between the resin's thermal properties conductivity and that of soil.

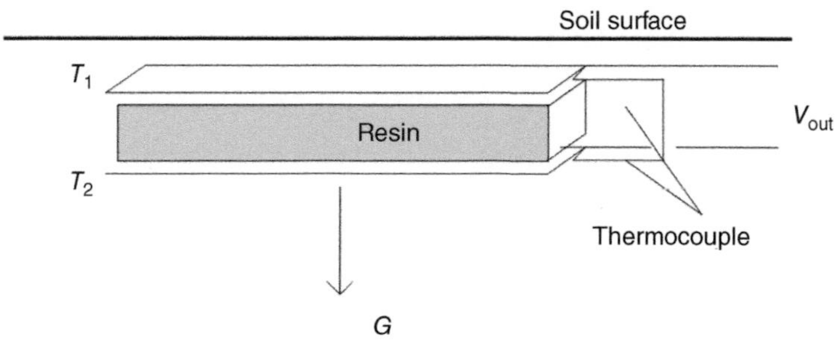

Figure 5.26 Principle of a soil heat flux plate, which determines the vertical heat flux density G by the temperature difference caused by the heat flow through a layer of resin of known conductivity.

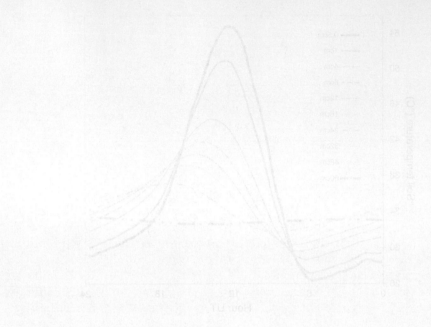

6

Humidity

The measurement of water vapour present in air – *hygrometry* – allows evaluation of atmospheric changes associated with precipitation and heat transfer between the surface and the atmosphere. Humidity is also an important quantity in other systems, for example in determining moisture loss from biological systems or even human comfort, and in influencing the propagation of radio waves through the atmosphere.

Humidity can be evaluated in a variety of ways, largely related to the different measurement techniques and so conversions to allow comparison between different instruments require some knowledge of the different quantities concerned. The physical measures are based on water vapour's different properties as a gas (vapour pressure), as a mixture (absolute, specific and relative humidity) and by the thermodynamic change associated with the presence of water vapour (wet bulb and dew point temperatures).

6.1 Water vapour as a gas

The relationship between the different humidity quantities can be appreciated by considering a thermally isolated container of fixed volume V partly filled with water at a temperature T (Figure 6.1).

The pressure due to the water vapour in Figure 6.1, considered as a gas, offers one way of characterising the humidity. If there is a mass m of water vapour present in the container of volume V at a temperature T, the pressure exerted by the vapour can be found from the ideal gas law[i] as

$$eV = nR^*T \,, \tag{6.1}$$

for n the number of moles of water present and R^* the universal gas constant.[ii] Expressing this in terms of the mass of water vapour m, the vapour pressure is

$$e = \left(\frac{m}{V}\right)\frac{1}{M_r}R^*T \,, \tag{6.2}$$

[i] Water vapour at low concentration can be regarded as an ideal gas, and obeys Dalton's law of partial pressures.

[ii] Note that in meteorological usage, R is often used to signify the gas constant for dry air rather than the universal gas constant. For clarity, the universal gas constant is written here as R^* (= 8.314 J K^{-1} mol^{-1}).

Meteorological Measurements and Instrumentation, First Edition. R. Giles Harrison.
© 2015 John Wiley & Sons, Ltd. Published 2015 by John Wiley & Sons, Ltd.
Companion website: www.wiley.com/go/harrison/meteorologicalinstruments

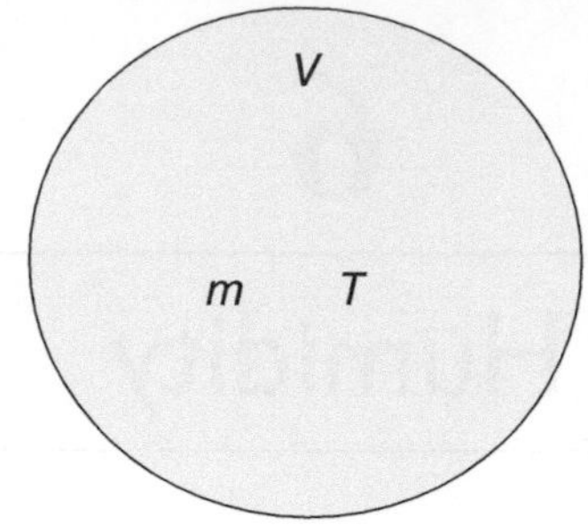

Figure 6.1 A container, of volume V and at a temperature T, in which there is a mass of water vapour m.

where M_r is the relative molecular mass of water. The term in the brackets on the right-hand side can be seen to be the mass per unit volume of the moist air, which is a measure of concentration also known as the absolute humidity (see also Section 6.2.1). Hence, at constant temperature, the vapour pressure e is proportional to absolute humidity.

If the container of Figure 6.1 has further liquid water added and all other conditions remain the same, the amount of vapour present will change as increased exchange of water molecules from liquid to vapour and vapour to liquid occurs. When these two exchange processes occur at equal rates, equilibrium is reached and the vapour is said to be saturated with respect to liquid water (Figure 6.2). The vapour pressure in this saturated condition can be calculated from Equation 6.2 for the saturated mass of water present m_s, to find the saturation vapour pressure e_s.

It is found that the saturation vapour pressure e_s varies with temperature. This means that for any given temperature T, there is an associated maximum value of the saturation vapour pressure e_s. The variation of e_s with temperature $e_s(T)$ is described by the Clausius–Clapeyron equation, which considers the equilibrium between two phases of the same substance. The Clausius–Clapeyron equation says that

$$\frac{1}{e_s}\frac{de_s}{dT} = \frac{M_r\lambda}{R^*T^2}, \tag{6.3}$$

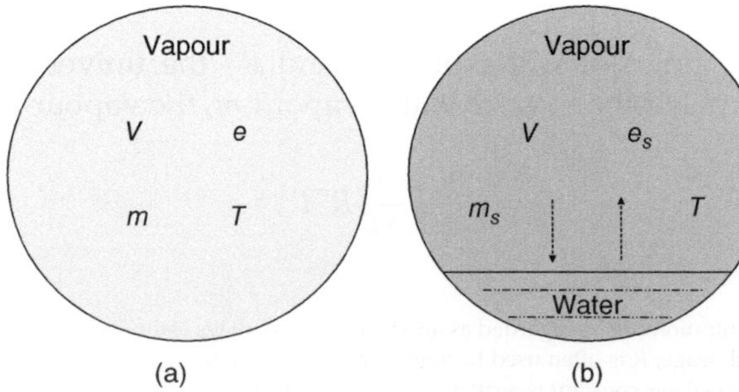

Figure 6.2 Comparison between a container in which (a) some water vapour is present and (b) one in which liquid water is also present.

Table 6.1 Coefficients for calculating the saturation vapour pressure over liquid water and ice using Equation 6.4

	Water (Bolton's coefficients)	Ice
A (hPa)	6.112	6.109
B	17.67	22.50
C (°C)	243.5	273.0
Range	−35°C to 30°C	−50°C to 0°C

The coefficients for liquid water are due to Bolton [58] and those for ice have been obtained by a fitting to the calculated values of Reference 59.

where λ is the latent heat of vaporisation of water. The form of the function $e_s(T)$ implied is important in meteorology because of the effects on condensation across the range of temperatures encountered. This temperature variation can, in principle, be found by integration of Equation 6.3, but, for accurate work, allowance is also needed for the variation of λ with temperature. This leads to many different empirical formulae for calculating $e_s(T)$, the most accurate of which use multiple term polynomials to represent the available experimental data. One much simpler parameterisation, often, but incompletely, solely [57] attributed to Magnus,[iii] is

$$e_s(T_C) = A \exp\left[\frac{BT_C}{T_C + C}\right], \quad (6.4)$$

for which empirically derived coefficients *A*, *B* and *C* for liquid water are given in Table 6.1 for T_C in Celsius.

A similar representation can be applied to supercooled water (water which remains liquid below its usual melting temperature). In comparison with supercooled water at the same temperature, the saturation vapour pressure over ice is slightly lower, which means that air saturated with respect to liquid water will be supersaturated with respect to ice. The saturation vapour pressure with respect to ice can be calculated using the same simplified form used for water, using slightly different empirically derived coefficients; these are also given in Table 6.1. Values of the saturation vapour pressure calculated with the simplified forms for ice and water are compared with those found using full polynomial calculations in Figure 6.3.

6.2 Physical measures of humidity

Although the vapour pressure due to the water present provides a fundamental measurement of humidity, practical instrumentation is varied and employs a range of hygrometric properties of materials. These include quantities which are related to the mixture of air and moisture (absolute, specific and relative humidity) and the thermodynamic changes which occur as a result of the presence of water vapour (dew point

[iii] Magnus' nineteenth century experimental work built on earlier experiments of Dalton, and the theoretical work of Roche and August, in arriving at a useful formula. Tetens subsequently provided modified coefficients in 1930.

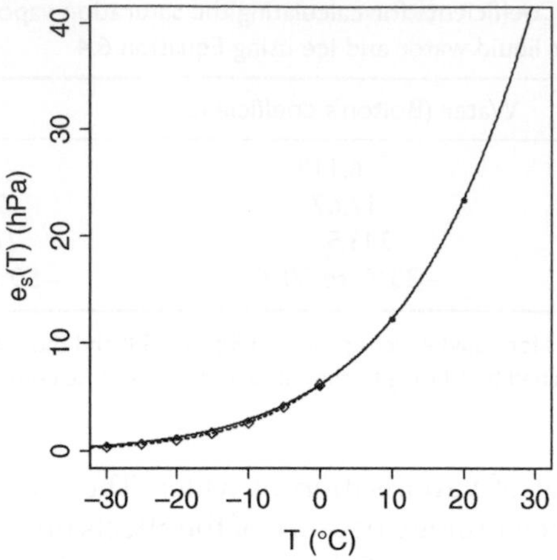

Figure 6.3 Saturation vapour pressure e_s over liquid water (points) and ice (diamonds) at temperature T, calculated from an accurate polynomial fit [59] to experimental data. Solid and dashed lines show calculations made using Equation 6.4 employing the coefficients of Table 6.1, for liquid water and ice respectively.

temperature and wet bulb temperature). The different quantities are now briefly considered and their interchange by calculation summarised in Table 6.2, before examining the instruments used to make the measurements.

6.2.1 *Absolute humidity*

As apparent from Equation 6.2, the vapour pressure is proportional to the mass of water vapour per unit volume of moist air (i.e. the volume of air containing the water vapour). This mass per unit volume is effectively a concentration of water vapour, which is known as the absolute humidity χ. From Equation 6.2, χ can be defined in terms of vapour pressure e and absolute temperature T as

$$\chi = \frac{M_w}{R^*}\frac{e}{T}, \tag{6.5}$$

where M_w is the relative molecular mass of water.

6.2.2 *Specific humidity*

An alternative method of expressing the humidity is to express the mass of vapour as a fraction of the mass of the moist air, rather than its volume. The specific humidity of a sample of moist air is defined as

$$q = \frac{m_v}{m_d + m_v}, \tag{6.6}$$

with m_v is the mass of water vapour present and m_d the mass of the dry air.[iv] Considering a unit volume of moist air at air pressure p with water vapour pressure e, this becomes

$$q = \frac{\frac{eM_w}{R^*T}}{\frac{pM_a}{R^*T} + \frac{eM_w}{R^*T}} \approx \frac{M_w}{M_a}\frac{e}{p}, \tag{6.7}$$

where M_w is the molecular mass of water (18 g mol^{-1}), M_a the molecular mass of air (29 g mol^{-1}) and T is the absolute temperature. The approximation in Equation 6.7 is good (to about 1%), because the contribution of the water vapour to the total mass of the moist air is small. Specific humidity is also sometimes expressed in terms of the humidity mixing ratio r_v (the ratio of the mass of vapour to the mass of dry air) as

$$q = \frac{r_v}{1 + r_v}. \tag{6.8}$$

6.2.3 *Relative humidity*

The saturation vapour pressure e_s and vapour pressure e give rise to the concept of relative humidity, which is also apparent in the comparison of Figure 6.2 between sub-saturated and saturated equilibrium conditions. In Figure 6.2a, the vapour pressure e is less than the saturation vapour pressure e_s of Figure 6.2b. The relative humidity of an air sample is the ratio of the actual vapour pressure e at a temperature T to the maximum value of vapour pressure (the saturation vapour pressure, e_s) at the same temperature. The relative humidity RH is defined by the percentage

$$RH = 100\frac{e(T)}{e_s(T)}. \tag{6.9}$$

6.2.4 *Dew point and wet bulb temperature*

The dew point temperature T_{dew} is the temperature at which air just becomes saturated with respect to a plane surface of pure water, when it is cooled from a higher temperature T at constant pressure. This turns out to have practical relevance, as the dew point can be detected optically when condensation begins to appear after cooling (see Section 6.3.2). Since no water vapour is added or removed during the cooling, the vapour pressure does not change and

$$e(T) = e_s(T_{dew}). \tag{6.10}$$

[iv] The specific humidity is dimensionless, but it is often expressed in g kg^{-1}.

Alternatively, the relative humidity *RH* and the dew point temperature are related by

$$RH = 100\frac{e_s(T_{dew})}{e_s(T)} . \tag{6.11}$$

Hence, if the relative humidity is known, the dew point temperature can be found from Equation 6.11. Using the approximation of $e_s(T)$ from Equation 6.4, solving for T_{dew} gives

$$T_{dew} = \frac{C\left[\ln\left(\frac{RH}{100}\right) + \left(\frac{BT}{T+C}\right)\right]}{B - \ln\left(\frac{RH}{100}\right) - \left(\frac{BT}{T+C}\right)} . \tag{6.12}$$

where *B* and *C* are from Table 6.1 for water vapour and *T* is in Celsius.

A quantity related to the dew point is the wet bulb temperature T_w, which is the temperature of air into which additional water vapour has been evaporated isobarically, causing cooling until the air becomes saturated. This quantity is closely associated with the psychrometer (see Section 6.3.7), which determines T_w directly. In unsaturated conditions, T_w is greater than T_{dew} and simultaneous variations in relative humidity, wet bulb and dew point temperatures during a summer day are shown in Figure 6.4.

Calculations for the conversion between some of the different measures of humidity are summarised in Table 6.2.

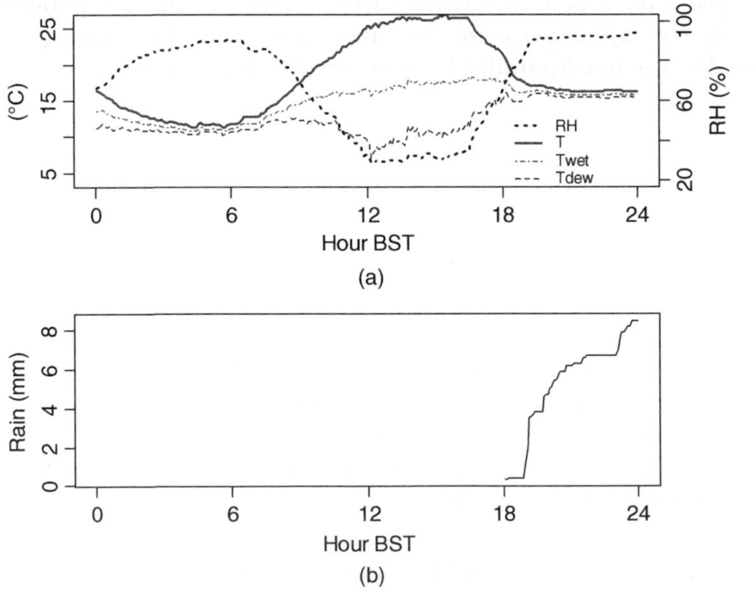

Figure 6.4 Variations in (a) humidity variables (Relative Humidity, air temperature, wet bulb temperature and dew point temperature) and (b) rainfall, on a summer day (27 July 2013) ending in a shower. (The relative humidity was measured independently of the wet bulb and dew point temperature.)

Table 6.2 Numerical conversions between different measures of humidity

Quantity		Remarks
Specific humidity q	$q = 0.62\frac{e}{p}$	q in g kg^{-1}; units of vapour pressure e and air pressure p consistent
Absolute humidity χ	$\chi = 216.7\frac{e}{T_K}$	χ in g m^{-3}; vapour pressure e in hPa and air temperature T_K in Kelvin
Relative humidity RH	$RH = 100\frac{e}{e_s(T)}$	RH expressed as a percentage, from vapour pressure e and saturation vapour pressure at temperature T, with consistent pressure units
Dew point temperature T_{dew}	$T_{dew} = \frac{243.5\left[\ln\left(\frac{RH}{100}\right) + \left(\frac{17.67\,T_C}{T_C+243.5}\right)\right]}{17.67 - \ln\left(\frac{RH}{100}\right) - \left(\frac{17.67\,T_C}{T_C+243.5}\right)}$	Air temperature T_C and dew point temperature in Celsius

6.3 Hygrometers and their operating principles

6.3.1 *Mechanical*

Some natural substances are sensitive to moisture, which can be exploited for humidity measurements. Humidity modifies the mechanical properties of hair, goldbeater's skin[v] and animal horn or antler which are hygroscopic. For example, hair increases its length by about 2% as RH increases from 0% to 100%, although the length change is only approximately linear, decreasing at large RH.

In a hair hygrometer, the change in length of a hair is amplified mechanically and displayed on a scale, in some cases with a correction applied. Such devices are simple in principle and require no power, but have poor accuracy (±5% RH at best), and poor stability, needing frequent recalibration. Mechanical sensing methods have been used for upper air measurements in radiosondes, but, as a hair sensing element is sensitive to dirt and dust, it is more suitable for use within autographic instruments (hygrographs) used to record trends, rather than to maintain measurement stability over long periods. A hair hygrograph may also be combined with a thermograph (Figure 6.5).

The non-linear response of a mechanical hygrometric element to humidity can be partially compensated in some instruments, using linkages and levers to provide non-linear amplification. The principle of this is shown in Figure 6.6, although the detailed implementation varies from device to device.

In this configuration of hair hygrometer, the hair (DE) is stretched and coupled to the pointer via a hook (ABC), which is free to pivot about B. This yields a non-linear response of the pointer, proportional to the product of the cosines of the stretching angle θ and pivoting angle α. At high humidity $\cos\alpha > \cos\theta$; this approximately compensates for the reduction in the hair sensitivity with increasing RH. An example of the tension used on the hair in such devices is given in Figure 6.7.

[v] Goldbeater's skin is a parchment used in making gold leaf. It is made from chemically treated cattle intestine, stretched and beaten flat and thin.

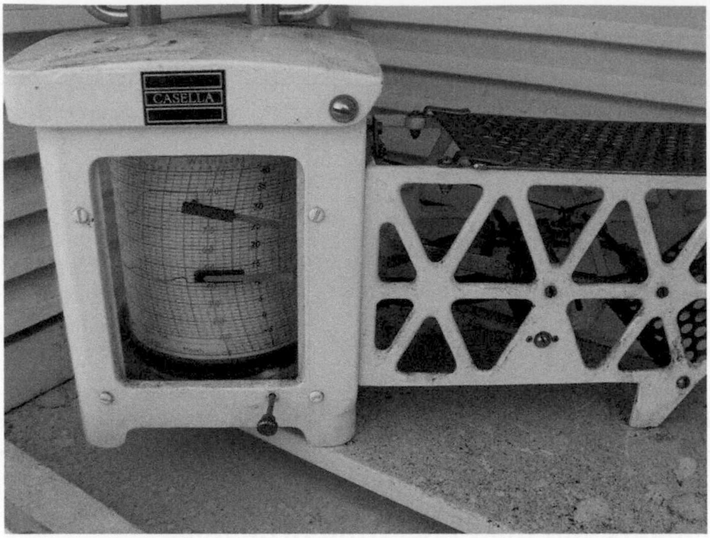

Figure 6.5 Thermohygrograph, for continuous recording of changes in temperature and relative humidity.

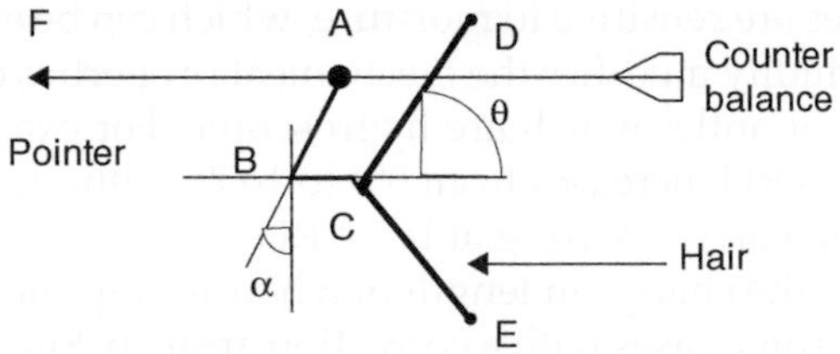

Figure 6.6 Schematic of the mechanical linkage used in a hair hygrometer.

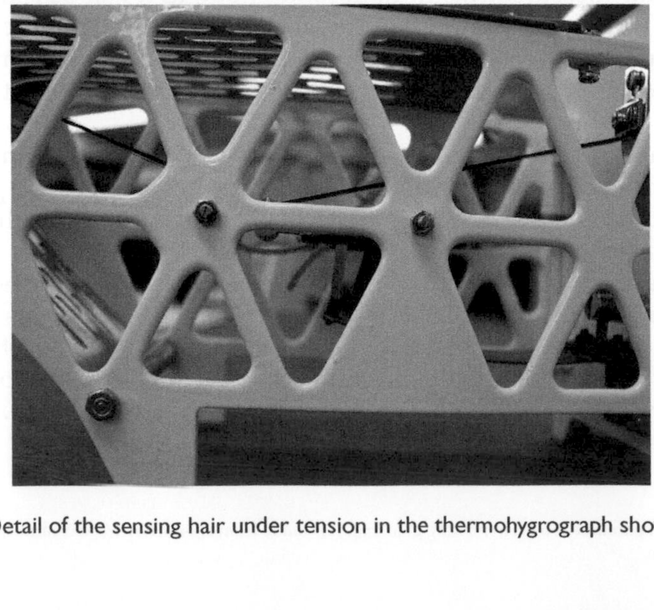

Figure 6.7 Detail of the sensing hair under tension in the thermohygrograph shown in Figure 6.5.

6.3.2 Chemical

The amount of water vapour present in an air sample can be measured by pumping the moist air through a pre-dried desiccant material. If the volume flow rate Q is known from a gas meter and the change in mass Δm of the desiccant measured during an experiment of duration t, the absolute humidity of the sample can be obtained directly as

$$\chi = \frac{\Delta m}{Qt} . \tag{6.13}$$

In principle, this is an absolute method, but uncertainties will arise if the desiccant becomes saturated, or if the change in mass is insufficient for it to be measured accurately. At low humidity, the need for a measurable change in mass dominating over all other changes may require an impractically long experimental duration, during which, if atmospheric air is sampled, the humidity may vary.

6.3.3 Electronic

In common with the mechanical sensors, electronic sensors can also respond to relative humidity changes. These sensors respond through a change in resistance (known as humidity sensing resistors or *hygristors*), or a change in capacitance (sometimes known by their tradename, *humicaps*).[vi] Hygristors are usually constructed from carbon particles within a hygroscopic film, with electrodes manufactured along the sides of the film. The resistance between the electrodes increases greatly with humidity as the film dimension changes and the carbon particles move apart. The resistance-RH sensitivity is highly non-linear, with a much reduced response at low humidity [60].

A humidity-sensing capacitor is specially constructed to give a change in capacitance with relative humidity. Capacitors consist of two large area electrodes, which are separated by an insulating dielectric material. The geometry of the electrodes, their separation and the dielectric constant of the separating material defines the device's capacitance (Figure 6.8). In the case of a humidity sensor, the dielectric is designed to be porous and to adsorb or release water vapour. Because water has a large dielectric constant, the capacitance of the device becomes sensitive to the ambient water vapour concentration within the dielectric.[vii]

The capacitance sensor shows a more linear response to humidity than for a hygristor, but its accuracy diminishes at large values of relative humidity close to saturation, such as in fog or cloud when it may also exhibit hysteresis. Capacitance sensors also have a relatively slow response time (~10 s to 100 s), which lengthens with decreasing temperature, and introduces a difficulty with using such sensors on radiosondes at altitude (see Section 11.2.2). The capacitance can be found by making it the frequency-determining element of an oscillator circuit (see Section 3.6.1), and measuring the frequency. A humidity-sensing capacitor is shown in Figure 6.9; for long term use, a micropore filter would be added for protection from dust and salt.

[vi] Humicap is a trademark of Vaisala.

[vii] Some designs of soil moisture sensor operate on the same principle of determining the dielectric constant, although the measuring capacitance is formed using electrodes of pointed tubular conductors, able to be driven into the soil.

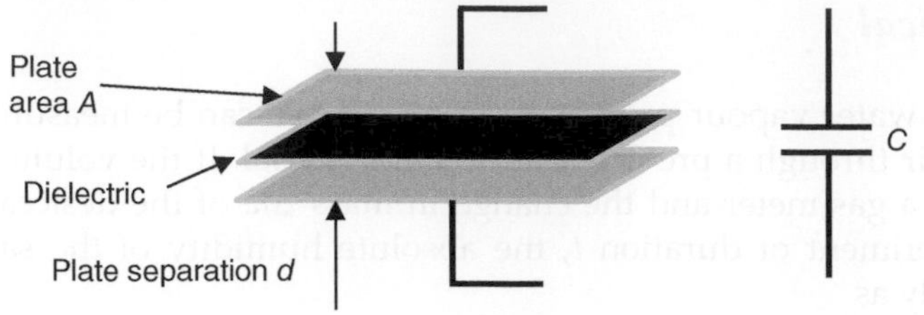

Figure 6.8 Construction of a parallel plate capacitor. The capacitance C is given by $C = \frac{\varepsilon_0 \varepsilon_r A}{d}$, where A is the electrode area, d the plate separation, ε_0 the permittivity of free space and ε_r the relative permittivity (dielectric constant) of the insulating material separating the plates.

6.3.4 Spectroscopic

Absorption of radiation may be used to measure specific humidity. The wavelengths used are typically infrared [61] at wavelength of 1 to 2 μm, or the shorter wavelength Lyman-α radiation (121.6 nm) which shows a stronger response [62]. The primary advantage of the absorption approach is the short response time, and consequently, rapid fluctuations in humidity can be observed. The disadvantage is that such instruments do not show good long-term stability, due to variations in the radiation source and detector used and the stability of filters employed. This does not cause a problem if only the fluctuations are required, but a further independent measurement technique is required if absolute results are needed.

Figure 6.10 shows the principle of operation of an infrared hygrometer which passes radiation through air using a short path. The path is typically about 10 to 20 cm in length, and may be folded to increase the sampling length using a mirror.

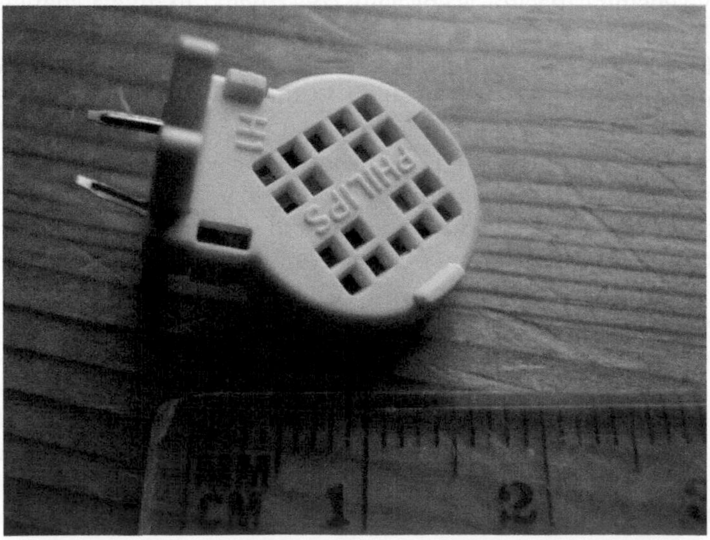

Figure 6.9 Example of a humidity-sensing capacitor. The outer case is porous to allow free exchange of ambient air with a non-conducting foil dielectric able to adsorb water vapour, which is gold-coated on each side.

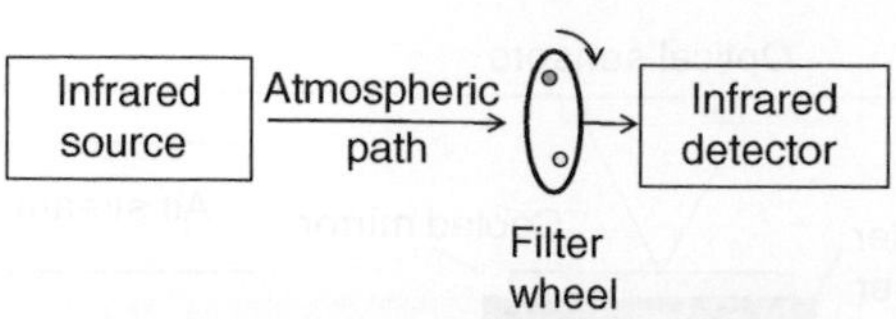

Figure 6.10 Operating principle of an infrared hygrometer. An infrared source provides broadband infrared radiation, including one wavelength at which water vapour absorbs and one at which it does not. These wavelengths are selected by tuned optical filters, placed, alternately in front of the infrared detector at a rate to which subsequent signal processing circuitry is synchronised.

The source is broadband, in that the wavelengths it generates are not solely at a single wavelength, but includes one wavelength which is absorbed by atmospheric water vapour and another wavelength which is not. Filament lamps run at low currents provide suitable radiation sources. The two wavelengths chosen are measured separately by the same infrared receiver synchronised to a rotating chopper wheel. This chopper wheel carries optical filters tuned to the two different wavelengths,[viii] to provide two measurements, alternately, of solely the source variations, and of the source variations with the added effect of atmospheric absorption. By combining the two measurements made alternately and averaged, the effect of source variations can be reduced. This allows the variations due to the atmospheric water vapour fluctuations to be extracted despite the presence of other variations.

(Figure 12.2c shows an example of rapid response specific humidity measurements obtained using an infrared hygrometer.)

6.3.5 *Radio refractive index*

Propagation of very high frequency (vhf) radio waves in the atmosphere is affected by refraction. The refractive index n of air defines the propagation speed v of the radio waves, as

$$v = \frac{c}{n}, \tag{6.14}$$

where c is the speed of light in a vacuum. n is about 1.0003 at the earth's surface, so it is more conveniently expressed in terms of refractivity N, defined as $N = (n - 1) \times 10^6$. The refractivity is related to air temperature T, water vapour pressure e and total pressure P by

$$N = \frac{77.6}{T}\left(P + \frac{4810e}{T}\right). \tag{6.15}$$

This equation for refractivity thus essentially consists of a 'dry' term concerning air pressure, and a 'moist term' concerning the water vapour pressure. Measurement of the local atmospheric refractivity, for example using a microwave cell to determine v

---

[viii] By the use of a range of wavelengths, the same infrared absorption device can be used to measure other trace gases present such as CO_2 or CH_4. A device operating on this principle is also known as an IRGA (*InfraRed Gas Analyser*).

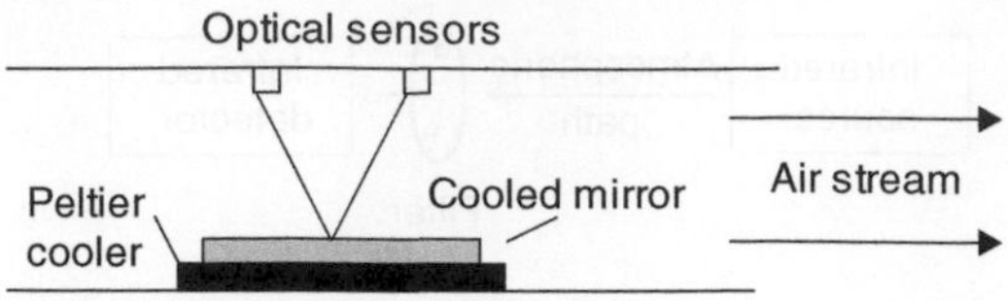

Figure 6.11 Schematic of a dew point meter using a Peltier cooling system to determine the temperature at which dew just forms on a cooled mirror by an optical detection technique.

or from weather radar transmissions, may therefore be used to provide information on the atmosphere's spatial and temporal humidity structure [63].

6.3.6 *Dew point meter*

The measurementof dew point temperature requires cooling of sub-saturated air until condensation occurs (Equation 6.10). A dew point meter implements this cooling, condensation detection, and the determination of inlet and condensation temperatures. In a dew point meter, air is passed slowly over a small mirror cooled electrically by a Peltier device, until dew just forms on the surface, when the mirror's surface temperature is found (Figure 6.11). This temperature measurement is effectively adjacent to the cooled air, and therefore closely approximates T_{dew}. The dew detection can be made automatic by using optical reflection from the cooled surface and a light beam, which allows continuous automatic temperature control and therefore continuous determination of T_{dew}. The time response depends on the effectiveness of the cooling system for the size of mirror employed, but can be made to operate at timescales around 1 s by miniaturising the mirror and optical detection system. Maintaining cleanliness of the mirror is, however, a practical limitation.

6.3.7 *Psychrometer*

A psychrometer consists of two matched thermometers of any kind (see Figure 6.12), one of which measures air temperature T (called the 'dry bulb') and the other thermometer is covered with wetted cotton (muslin) cloth and reads the 'wet bulb' temperature T_w. If little water vapour can be evaporated into the air around the wet bulb, there will be little cooling, whereas if considerable water vapour can be evaporated, there can be appreciable cooling. The difference between the wet bulb and dry bulb temperatures – known as the *wet bulb depression* – is therefore inversely proportional to the humidity.

The wet bulb temperature is the temperature of an air sample which has been cooled by evaporating water into it until it becomes saturated, with the latent heat required for this evaporation supplied from the internal energy of the air. Hygrometric tables can be used to calculate e or relative humidity from the dry and wet bulb temperatures. The principle on which such calculations are based is to relate change in temperature of the air to the amount of evaporation occurring. Evaporation acts to cool the air from the dry bulb temperature to a new temperature T_{wet} as shown in Figure 6.13.

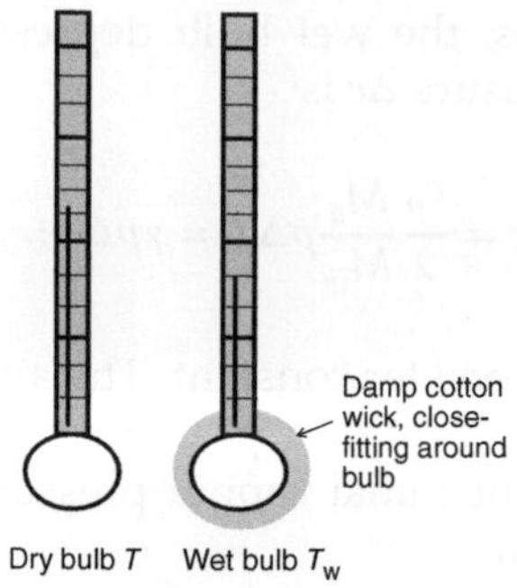

Figure 6.12 Schematic of a psychrometer constructed from two liquid-in-glass thermometers. A moistened cotton wick is fitted to one of the thermometers – the wet bulb thermometer – which records a lower temperature than the unmodified (dry bulb) thermometer if evaporative cooling is able to occur.

Consider the effect of evaporating a mass of water Δm into a mass M of moist air. From Equation 6.7, the change in vapour pressure Δe and Δm are related by

$$\frac{\Delta m}{M} = \frac{M_w}{M_a}\frac{\Delta e}{p}, \tag{6.16}$$

and, at constant pressure, the associated change in temperature ΔT results from the supply of the latent heat from the internal energy of the air,

$$\lambda \Delta m = c_p M \Delta T . \tag{6.17}$$

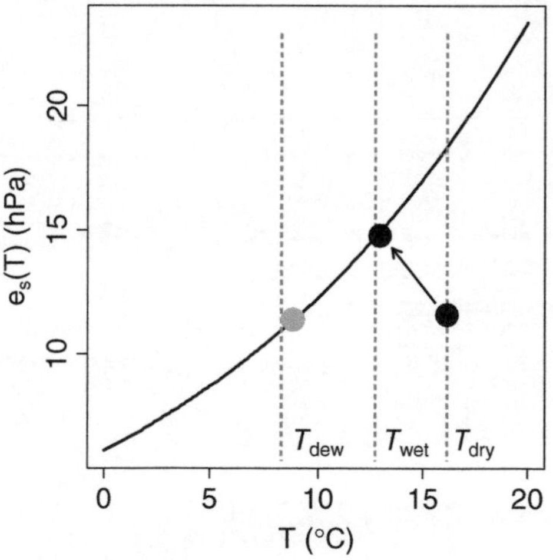

Figure 6.13 The cooling effect of evaporating additional water into air initially at a temperature T_{dry}, shown on a plot of saturation vapour pressure of water against temperature. Water added by evaporation from a wet bulb increases the vapour pressure, causing a change in position on the plot from the initial T_{dry} position, to a lower temperature T_{wet} with an associated increase in vapour pressure. (If the sample were just cooled, without adding water vapour, the effect would be just to move left along the temperature axis to T_{dew}, keeping constant vapour pressure, to the grey point.)

Combining these expressions, the wet bulb depression ΔT associated with an increase in the local vapour pressure Δe is

$$\Delta e = \frac{c_p}{\lambda} \frac{M_a}{M_w} p \Delta T = \gamma p \Delta T \, , \tag{6.18}$$

where γ is known as the psychrometer constant. The air is saturated at the wet bulb temperature, and the difference in vapour pressure resulting from the evaporation (evident in Figure 6.13) allows the initial vapour pressure to be calculated, using the so-called psychrometer equation

$$e = e_s(T_w) - \gamma p(T - T_w) \, . \tag{6.19}$$

6.4 Practical psychrometers

The operation of real psychrometers differs from the theoretical concepts outlined above, because of the need to ensure the thermometers measure accurately through good ventilation and protection from direct solar radiation. The ventilation acts both to reduce the radiation error, but, more importantly, also establishes consistent evaporation conditions. One design of practical psychrometer is the Assmann psychrometer, which contains the two thermometers in radiation shields, both of which are ventilated by a clockwork fan mechanism (Figure 6.14). Another implementation is

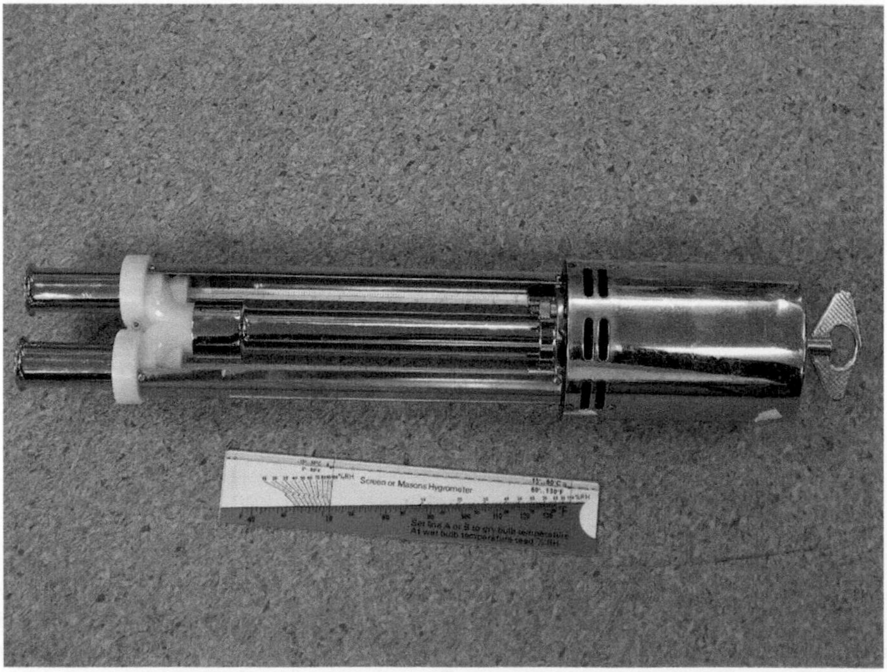

Figure 6.14 An Assmann psychrometer. This uses polished metal shields to protect the thermometers from radiation, and applies forced ventilation by a clockwork motor.

the whirling psychrometer, which holds the two thermometers in a frame, rotated by hand to provide forced ventilation.

For a real instrument, Equation 6.19 does not apply exactly to the wet bulb temperatures measured as the ideal value of the psychrometer constant is not reached. Instead, the vapour pressure can be calculated using the semi-empirical equation

$$e = e_s\,(T_w) - Ap\,(T - T_w)\ , \tag{6.20}$$

where A is an empirical psychrometer coefficient determined for a particular instrument. For force-ventilated psychrometer when $T_w > 0°\mathrm{C}$ with p in hPa, and A in K^{-1},

$$Ap = 0.667p/1000\ . \tag{6.21}$$

Hygrometric tables, slide rules (e.g. Figure 6.14), graphical representations (e.g. Figure 6.19) and computer programs are often used to make the calculation straightforward.

In a screen psychrometer the two thermometers are housed within a Stevenson screen (Figure 6.15), and the operation depends on sufficient natural ventilation occurring through the louvered sides. For such a naturally ventilated screen psychrometer, it is usually assumed that

$$Ap = 0.80p/1000\ . \tag{6.22}$$

However, the effect of insufficient ventilation on a screen psychrometer is to change the effective pyschrometric coefficient, which is considered further in Section 6.4.2.

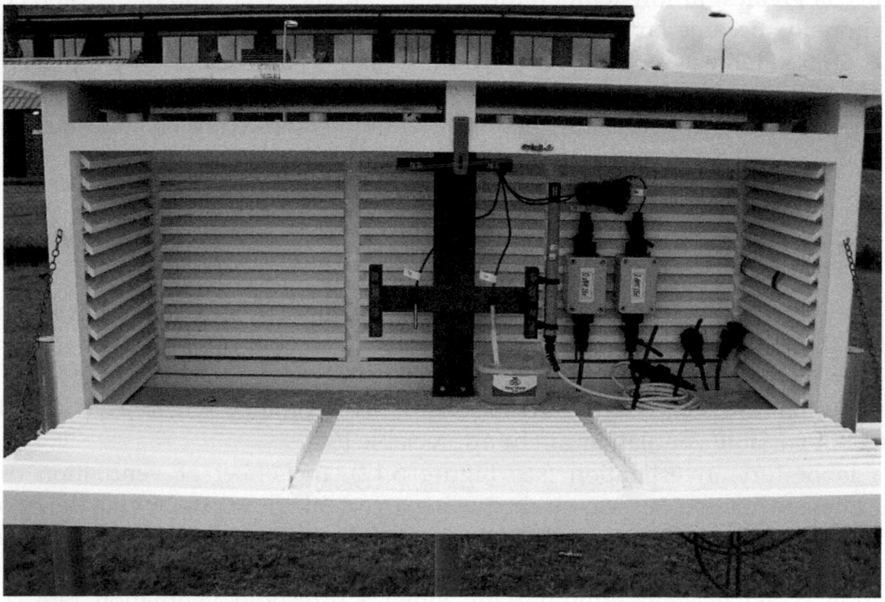

Figure 6.15 View of a psychrometer arrangement within a large thermometer screen. The two temperature measurements in the centre are obtained using platinum resistance sensors, the right one of which has a cotton wick attached to form a wet bulb thermometer.

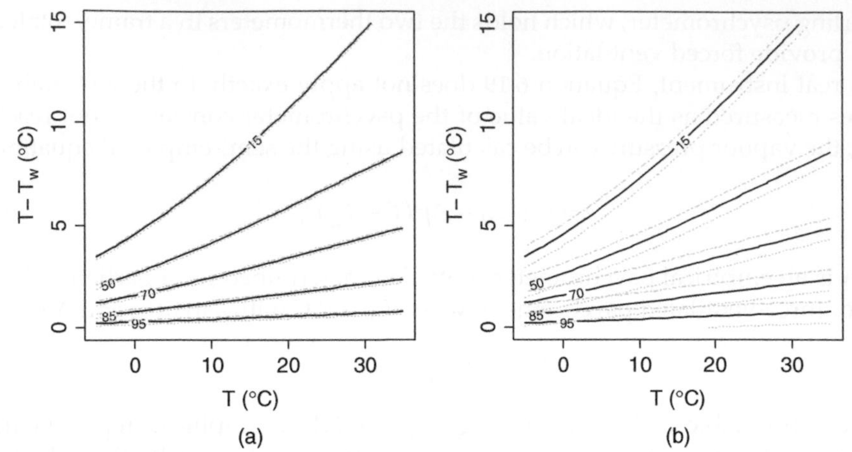

Figure 6.16 Effect of uncertainty in dry and wet bulb temperatures on the derived relative humidity (shown as contours of RH at 15%, 50%, 70%, 85% and 95%, with thin lines marking the range), calculated for a screen psychrometer (a) with temperature uncertainties of ±0.1°C, and (b) temperature uncertainties of ±0.5°C.

6.4.1 *Effect of temperature uncertainties*

The accuracy of vapour pressure determinations using a psychrometer strongly depends on the accuracy of the temperature measurements, and therefore the matching of the thermometers used. The effect of thermometer uncertainty on the derived humidity is greatest at low temperatures, as the smaller wet bulb depressions can, proportionally, become large. For example, for a 1°C variation in wet bulb temperature at 60% RH, the associated variation in RH is 17% at 0°C but only 6% at 30°C. Figure 6.16 illustrates the effect of temperature uncertainty on the calculated RH. In practice, it can also be difficult to keep the wet bulb consistently wet and the water clean.

6.4.2 *Ventilation effects*

Laboratory experiments show that the variations in ventilation rate also have a large effect on the humidity determined by a psychrometer. For a force ventilated psychrometer such as a whirling or Assmann device, the effect is small as long as a critical ventilation rate is exceeded, but, for a naturally ventilated screen psychrometer, the effect of variable ventilation can be appreciable [64].

In a laboratory investigation (see Figure 6.17), the effect of ventilation on a psychrometer arrangement was studied using a variable flow rate, with a dew point meter to provide reference values of humidity. Figure 6.18a shows, for ventilation speeds greater than about 3 m s^{-1}, A tends to a constant value, so the actual ventilation speed becomes unimportant above a threshold ventilation rate. This asymptotic variation provides the basis on which the vigorous clockwork ventilation for the Assmann psychrometer operates, and is the reason why the whirling psychrometer can operate effectively without an accurate determination of the ventilation speed.

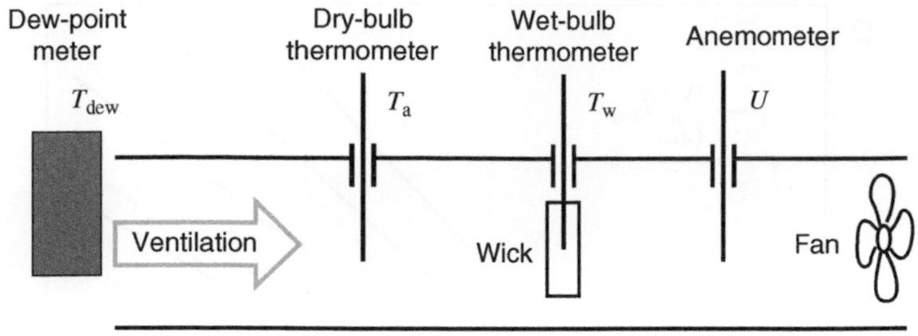

Figure 6.17 Apparatus to investigate the effect of ventilation on the psychrometer coefficient using electrical thermometers. The wind speed U is varied, and the incoming air's dew point T_{dew} and the dry and wet bulb temperatures T_a and T_w are measured when equilibrium is established.

For a psychrometer operating in a naturally ventilated Stevenson screen, the psychrometer coefficient must also be expected to vary. In particular, the standard value of psychrometer coefficient traditionally assumed (e.g. Equation 6.22, $A = 0.8 \times 10^{-3}\ \mathrm{K}^{-1}$) will be inadequate if ventilation is poor: field experiments [65] show $A = 1.2 \times 10^{-3}\ \mathrm{K}^{-1}$ is more appropriate for calm conditions (Figure 6.18b).

A non-linear model to describe the variation in effective screen psychrometer coefficient A_{eff} with wind speed u can be written in the form

$$A_{\mathrm{eff}}(u) = A_\infty + A_c \exp(-u/u_{\mathrm{min}}) , \tag{6.23}$$

which provides a finite value of A_{eff} when the wind speed u is zero. This allows the psychrometer coefficient to be corrected at low wind speeds. In this representation, A_∞ is effectively the well-ventilated (and therefore wind-speed

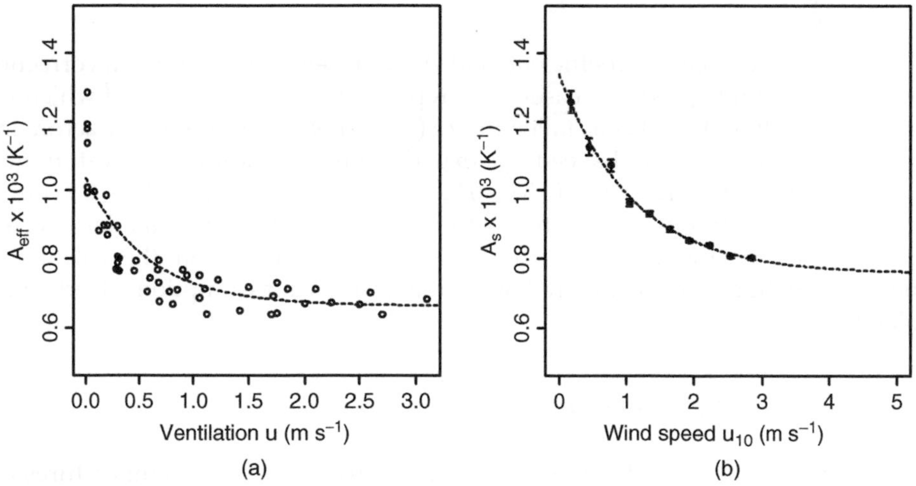

Figure 6.18 (a) Laboratory investigation of thermometer ventilation speed u on the effective psychrometer coefficient, A_{eff} using the apparatus of Figure 6.17. (b) Derived screen psychrometer coefficient A_s at Reading Observatory using a capacitance probe for the reference humidity measurements, plotted against wind speed measured at 10 m (u_{10}) (6816 hourly values for air temperatures between 5°C and 27.7°C were averaged into bins separated by 0.5 m s^{-1} in wind speed for $u_{10} < 3$ m s^{-1}; the standard error in the mean is shown).

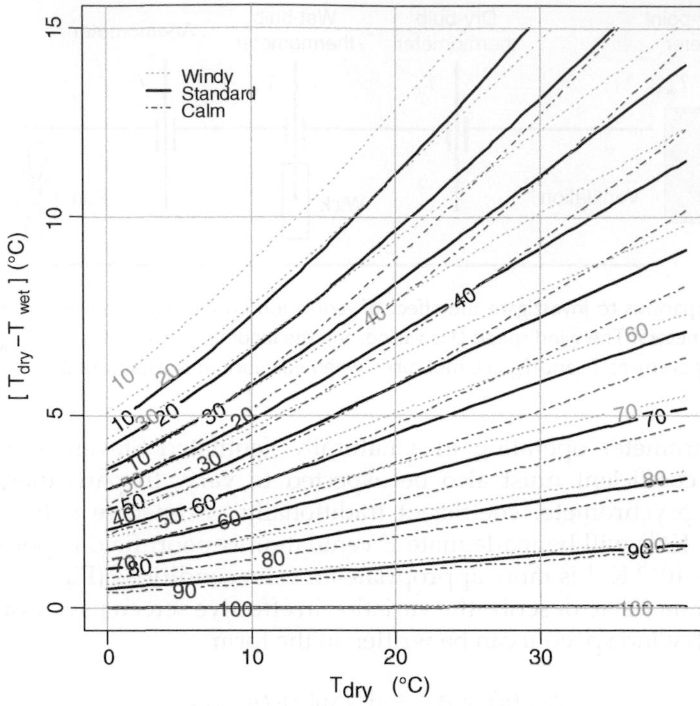

Figure 6.19 Effect of variation in ventilation on the Relative Humidity determined by a screen psychrometer, for given air temperature (x-axis) and wet bulb depression (y-axis). Solid lines show the relative humidity calculated using the assumption of the standard screen psychrometer coefficient; dotted and dot-dashed lines show, respectively, the relative humidity determined for windy and calm conditions, using the modified psychrometer coefficient of Equation 6.23.

independent) psychrometer coefficient, and A_c represents the additional correction required as the wind speed decreases. u_{min} represents the wind speed at which the correction contributed by A_c has fallen to $1/e$ (~63%) of that required for zero wind speed. For the 10-metre wind speed u_{10}, experiments at Reading Observatory give $A_\infty = (0.76 \pm 0.02) \times 10^{-3}\,K^{-1}$, $A_c = (0.58 \pm 0.02) \times 10^{-3}\,K^{-1}$ and $u_{min} = (1.1 \pm 0.1)\,m\,s^{-1}$, where the uncertainties in each coefficient represent one standard error. Figure 6.19 shows the effect on the ventilation correction on the calculation of the relative humidity, compared with the standard assumption of adequate ventilation.

6.4.3 *Freezing of the wet bulb*

Special precautions are needed in the use of psychrometers when temperatures are below zero. The wet bulb may be replaced by an 'ice bulb', for example established by removing the wick from the wet bulb, and dipping the thermometer in water. A thin layer of ice is then formed on the thermometer bulb, and the saturation vapour pressure is calculated for ice instead of liquid water, and A is again modified to a further empirical value ($0.594 \times 10^{-3}\,K^{-1}$).

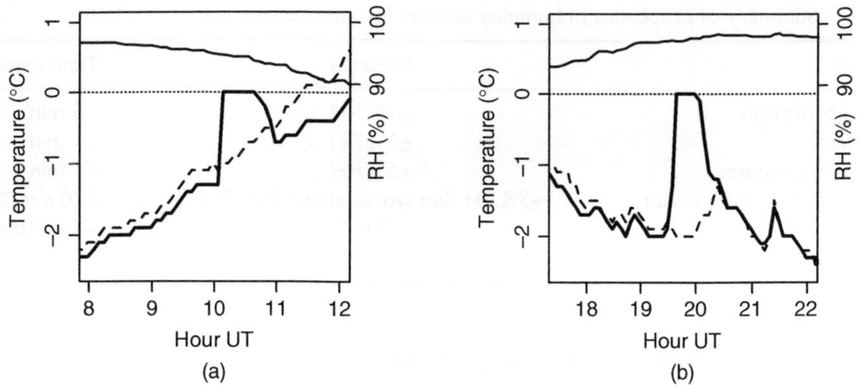

Figure 6.20 Sudden wet bulb temperature changes (thick black line) associated with sub-zero air dry bulb temperatures (dashed line), with capacitance probe relative humidity also shown (thin black line), for (a) day 331 and (b) day 339 of 2010, at Reading Observatory.

Establishing an ice bulb clearly requires intervention, but one interesting consequence from continuous measurements made by automatic logging apparatus is that the natural freezing of the screen psychrometer wet bulb can sometimes be observed, depending on the conditions at the site concerned. Examples are shown in Figure 6.20. Some supercooling of the wet bulb is evident in each case, and marked similarity in the wet bulb freezing event can be seen with the measurements of the droplet freezing in Figure 5.6, except for the timescales considered. Clearly, during the wet bulb freezing itself, the assumptions of the psychrometer theory become invalid.

6.5 Hygrometer calibration using salt solutions

To provide reference conditions of relative humidity for sensor calibration, one method is to sample the humidity above a saturated salt solution within a sealed container. These may take time to come to an equilibrium humidity, but different salts generate different humidity environments (see Table 6.3), some of which are remarkably temperature-stable [66].

Table 6.3 Relative humidity generated above saturated salt solutions (compiled using data from Philips Components [66] and Gregory and Rourke [67])

Salt		Relative Humidity (%) 20°C	25°C
Potassium nitrate	KNO_3		93.6
Potassium chloride	KCl	86.4	84.3
Magnesium nitrate	$Mg(NO_3)_2$		52.9
Sodium chloride	NaCl	76.5	75.3
Lithium chloride	LiCl	15	11.3

Table 6.4 Summary of properties of humidity sensors for atmospheric use

Instrument	Accuracy	Time response
Chemical absorption	±1% RH	~5 min
Psychrometer	±5% RH	~1 min
Mechanical hygrometer	±5% RH	~1 min
Electronic (capacitative) sensor	±3% RH (but worse above 90% RH)	~20 s to 2 min
Dew point meter	±0.1°C	~1 to 10 s
Absorption hygrometer	±10%	~0.05 s

6.6 Comparison of hygrometry techniques

In summary, fundamentally different humidity measurement techniques are available which are more or less appropriate in different situations. As an absolute determination of humidity, chemical absorption is accurate but it is inappropriate when the humidity is changing rapidly. For fluctuating conditions or turbulent measurements, spectroscopic absorption provides a very rapid response, but with poor absolute stability. A dew point meter can both be accurate and provide a response time of a few seconds. In terms of cost-effectiveness, the psychrometer provides reasonable accuracy if careful continuous attention is given to maintaining its condition and applying corrections if the ventilation is poor. Electronic sensors based on the capacitance method are robust and provide remote sensing possibilities, but their time response varies markedly with temperature, and their accuracy may be reduced as saturation is approached. A summary is given in Table 6.4.

7

Atmospheric Pressure

7.1 Introduction

Atmospheric pressure is a fundamental atmospheric quantity in providing synoptic information and because of its close relationship with height and atmospheric thickness. Accurate measurement of pressure altitude using sensors carried by radiosondes, in combination with temperature and wind, serve to define the dynamical state of the atmosphere.

The SI unit of pressure is the pascal, which is the pressure exerted by a force of 1 Newton acting on a perpendicular area of 1 m^2. Atmospheric pressure at mean sea level is approximately 10^5 Pa, or 1000 hPa, which is conveniently equal to 1000 mbar, an older unit of measurement. The desirable accuracy in pressure measurement at the surface or aloft is about ± 0.1 hPa, equivalent to 1 part in 10^4, which is a very demanding requirement for a routine scientific measurement. The radiosonde application (Chapter 11) is even more challenging, as the operating range of a radiosonde pressure sensor required may extend across roughly three orders of magnitude, during substantial changes in temperature.

Against this background of a demanding but routine measurement, surface pressure measurements are also conventionally corrected for the variation in pressure with height, which shows a reduction of about 1 hPa per 10 m of altitude near the surface. The pressure as observed without correction for altitude is known as the *station pressure*, whereas the *mean sea level pressure* is that corrected to the universally agreed sea level datum. In order that the correction itself does not degrade the measurement accuracy, the full sea level correction also allows for temperature effects (Section 7.3.1).

A measuring instrument for atmospheric pressure is known as a *barometer*, and one able to produce a chart record is known as a *barograph*.

7.2 Barometers

Instruments to measure atmospheric pressure have been used for over three centuries, and have been traditionally divided into two broad categories, depending on whether they use a liquid (usually mercury) as the sensing element, or not (*aneroid*

Meteorological Measurements and Instrumentation, First Edition. R. Giles Harrison.
© 2015 John Wiley & Sons, Ltd. Published 2015 by John Wiley & Sons, Ltd.
Companion website: www.wiley.com/go/harrison/meteorologicalinstruments

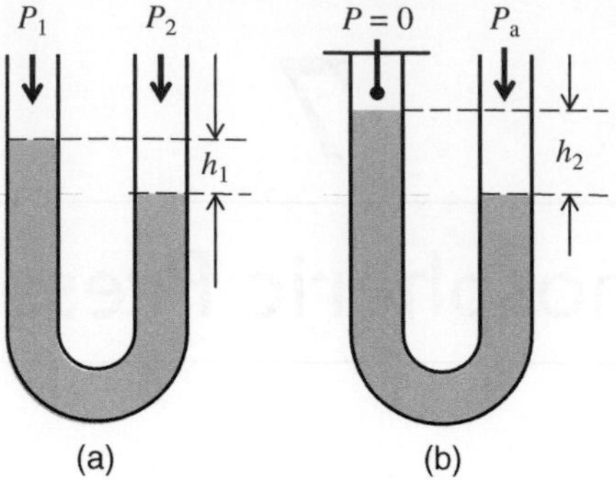

Figure 7.1 Pressure differences measured with a liquid manometer (a) for two finite pressures P_1 and P_2 ($P_2 > P_1$), giving a height difference h_1 and (b) between a vacuum ($P = 0$) and atmospheric pressure P_a giving a height difference h_2.

barometers).[i] Mercury barometers measure pressure by determining the height of a liquid column, which is related to pressure by the hydrostatic equation. Aneroid barometers use a thin metal chamber or diaphragm, having a membrane which deforms under pressure differences which can be measured mechanically or electronically. These are now discussed in turn, with the various techniques for measuring atmospheric pressure summarised in Table 7.1.

7.2.1 *Liquid barometers*

The position of a liquid in a tube is sensitive to the pressure difference to which the liquid is exposed. This means that pressure differences can be measured[ii] by the change in position of a liquid in a tube. If the tube is made symmetric, such as in a 'U' shaped tube, the difference in height of the fluid between one side and the other depends directly on the pressure difference applied (Figure 7.1) and the density of the liquid concerned.

The difference in height Δh is related to the pressure difference ΔP by the hydrostatic equation

$$\Delta P = \Delta h \rho g, \tag{7.1}$$

where ρ is the liquid's density and g the local gravitational force per unit mass (acceleration due to gravity). In the case illustrated by Figure 7.1a, the pressure difference and height difference are related by

$$P_2 - P_1 = h_1 \rho g, \tag{7.2}$$

[i] A further liquid-based indirect method of pressure measurement is that of the hypsometer, which determines height by determining the liquid's boiling point, when the variation of boiling point with atmospheric pressure is known (Section 7.2.3).

[ii] An instrument to measure pressure or pressure difference is a *manometer*. If one of the two pressures is the ambient pressure, the result is known as the *gauge pressure*, that is the difference from ambient pressure.

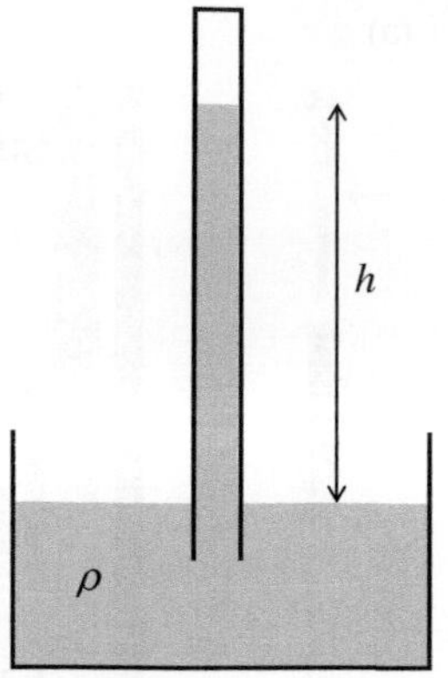

Figure 7.2 Principle of the liquid barometer. Accurate knowledge of the column height h, the liquid's density ρ, and the local acceleration due to gravity g are all required to determine the atmospheric pressure.

but, if the tube is sealed and evacuated, to establish a vacuum[iii] (Figure 7.1b), the atmospheric pressure and height vary directly together as

$$P_a = h_2 \rho g, \tag{7.3}$$

and hence the device measures absolute atmospheric pressure. This configuration is adapted to provide the operating principle of the liquid barometer, Figure 7.2.

As the measurement of atmospheric pressure using a liquid barometer requires knowledge of three other quantities, obtaining an accurate measurement of atmospheric pressure requires an accurate measurement of the column height h, the local acceleration due to gravity g and the liquid's density ρ. The fractional uncertainty in the calculated pressure $(\delta P/P)$ can be found by combining the individual fractional uncertainties as

$$\left(\frac{\delta P}{P}\right)^2 = \left(\frac{\delta h}{h}\right)^2 + \left(\frac{\delta \rho}{\rho}\right)^2 + \left(\frac{\delta g}{g}\right)^2, \tag{7.4}$$

(see also Section 2.2.2). This means that, to achieve $(\delta P/P) < 10^{-4}$, or 0.1 hPa in 1000 hPa as indicated in Section 7.1, each of the measurements of h, ρ and g must also be accurate to better than 1 part in 10^4.

7.2.2 *Mercury barometers*

Mercury is chosen for use in barometers because, as Equation 7.1 indicates, for ΔP of the order of surface atmospheric pressure, conveniently measurable values of Δh

[iii] The evacuated region in a mercury barometer is known as the Torricellian vacuum after the inventor of the barometer Evangelista Torricelli (1608–1647); the *Torr* remains used as a non-SI unit of pressure (1 hPa = 0.75 Torr). *Fortin* barometers include a reference setting for the mercury reservoir, originated by the instrument maker Jean Nicolas Fortin (1750–1831).

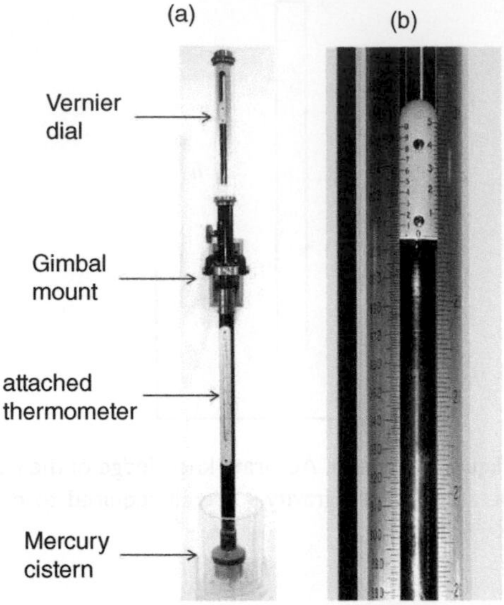

Figure 7.3 A Kew barometer, showing (a) the layout of the instrument and (b) the vernier dial used to determine the height accurately.

(<1 m) will require a dense liquid. In addition, mercury has a low vapour pressure at the temperatures required, and is easily cleaned. It does not wet the glass walls of tubes, which leads to formation of a convex meniscus, providing a well-defined measurement datum. A disadvantage is that mercury does show appreciable thermal expansion, and also mercury barometers are not very portable so are unsuitable for remote automatic data logging. They are usually used as free standing devices for station barometers, commonly arranged to hang vertically from a hook, or mounted on gimbals. A further notable difficulty is the considerable volume of liquid needed, which presents a hazard under failure conditions, as mercury is a cumulative poison, absorbed through the skin.

One commonly used practical mercury instrument for synoptic and climatological measurements is the Kew pattern barometer (Figure 7.3). This uses a mercury storage cistern and a vernier scale to find the height accurately. Air temperature T is measured with a thermometer attached to the barometer, to allow corrections to be applied.

There are some detailed practical considerations specific to mercury barometers. Surface tension changes occur in mercury barometers due to contamination of mercury and/or the barometer's tube which can be corrected by cleaning the barometer and mercury. A vacuum defect such as from mishandling is more serious, generally requiring refilling of the barometer and recalibration.

The Kew barometer is calibrated for use at an air temperature of 0°C and for standard value of g (9.80665 m s^{-2}). In practice, however, measurements will be made in circumstances which differ from these conditions and several elaborate corrections are necessary for the target accuracy required. These corrections allow for the influence of temperature on the density of mercury, the local value of g and imperfections in the barometer tube and scale, as tabulated by the manufacturer. For the Kew barometer [68], the corrections required are: (1) for expansion of the mercury and the scale,

$$P = \frac{g}{9.80665}\left\{B + i - B\left[\frac{(\alpha - \beta)T}{1 + \alpha T}\right] - f\frac{V}{A}(\alpha - 3\times 10^{-5})T\right\}, \tag{7.5}$$

where P is the corrected barometer and B the uncorrected reading, i the 'index' error due to errors in capillary, scales and vacuum), α the coefficient of expansion for mercury, β the linear expansion of the scale, T the instrument temperature, V the volume of mercury contained, A the effective area of the cistern, and f a unit conversion factor,[iv] and (2) for the local gravitational acceleration g, as

$$g = [9.80616\times(1 - 2.6373\times 10^{-3}\cos 2\varphi + 5.9\times 10^{-6}\cos^2 2\varphi)] - 3.086\times 10^{-4}h, \tag{7.6}$$

where g is the acceleration due to gravity (m s^{-2}) for a station which is at a latitude ϕ and a height h (metres) [69].

7.2.3 *Hypsometer*

A hypsometer is an instrument which primarily measures height, by determining the temperature at which a liquid boils. This can provide an indirect determination of pressure, as boiling occurs when the saturation vapour pressure of the liquid equals the local atmospheric pressure. Hence, by knowing the relationship between the saturation vapour pressure and temperature (such as Equation 6.4), and recording the liquid's boiling temperature, pressure can be found. The water hypsometer is important historically, as it provided one of the first techniques for an indirect measurement of altitude. Such an approach has also been used for pressure measurement in radiosondes (see Section 11.2.1), particularly as the sensitivity to pressure improves at lower pressures aloft, using liquids such as carbon disulphide or chlorofluorocarbons.

7.2.4 *Aneroid barometers*

Aneroid barometers are more convenient to use than mercury barometers and avoid the hazard associated with use of mercury. The sensor in an aneroid is a small metal capsule or diaphragm which is substantially evacuated, and of an elastic modulus to allow it to be distorted by atmospheric pressure changes. This change in dimension can be monitored mechanically or electrically (Figure 7.4), through changes in capacitance or inductance. Aneroid sensors can be made cheaply, and hence are widely used in domestic wall barometers, pressure altimeters and barographs. In these instruments, the displacement of the aneroid capsule is usually converted to a pointer's displacement by a mechanism of levers or pulleys. The properties of the mechanism limits accuracy of the dial reading, as it may stick or exhibit hysteresis.

[iv] These quantities are usually provided in a calibration certificate.

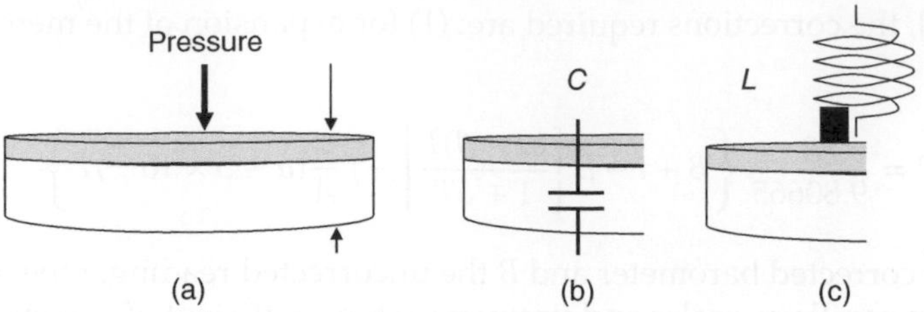

Figure 7.4 (a) Aneroid capsule concept in which vertical deformation occurs due to pressure changes. Electrical sensing methods can use the deformation change by (b) modifying the dimensions defining a capacitance *C* or (c) by displacing a permeable core within a coil of wire forming an inductance *L*.

In a barograph (Figure 7.5), multiple aneroid capsules are connected together to increase the sensitivity to pressure changes. The barograph is usually set up to measure the mean sea level pressure without correction, as, although not very accurate (±1 hPa), its importance is in providing a continuous record of pressure tendency.

7.2.5 *Precision aneroid barometers*

Precision aneroid barometers can be made robust enough for routine pressure measurements, and show little drift with time with no temperature correction required. The sensor in a precision aneroid barometer is a metal capsule, consisting of two thin diaphragms, separated by a flexible cylindrical metal bellows from which almost all the air has been removed. Pressure changes cause a relative movement between the diaphragms, which is measured by as a change in distance using an adjustable electrical contact moved on a fine micrometer screw thread. This is calibrated directly in terms of pressure.

Figure 7.5 A mechanically recording barometer (or barograph). The set of aneroid capsules connected together amplifies, with connecting levers, the change in response to pressure to cause an observable mechanical deflection on the recording chart drum.

In use, the screw thread is adjusted until the electrical contact is just broken, as shown by a visual indicator. Temperature compensation is achieved by leaving a small amount of gas in the capsule, and by the use of compensating metal linkages. Since the total relative movement of the diaphragms is only about 1.5 mm for the whole range of surface pressure (~900 to 1050 hPa), the measurement mechanism has to be very precisely engineered. In practice, several bellows connected in series may be used to increase sensitivity.

7.2.6 *Flexible diaphragm sensors*

These are based on a similar principle to that of an aneroid capsule, but rely either on detecting the flexing of a thin silicon diaphragm with changes in pressure, or changes in capacitance associated with changes in the separation between the diaphragm and a fixed plate. The latter type of sensor can be very small and light, and is used in radiosondes for measuring the pressure at the altitude reached by the instrument package.

An alternative approach is to couple the diaphragm to a resistive strain gauge, for a direct electrical output related to pressure. Such sensors are suitable for remote and automatic pressure measurements, but suffer from substantial temperature errors. Compensation systems, temperature characterisation or temperature stabilisation are therefore required [70].

7.2.7 *Vibrating cylinder barometer*

A vibrating cylinder barometer is capable of very high accuracy and stability. The operating principle is based on detecting the natural frequency of oscillation in the 'hoop' vibrating mode of a thin-walled cylinder (open to the atmosphere), which is surrounded by a near vacuum. The oscillations are excited by magnetic forces from electromagnetic coils placed around the cylinder and detected by the induced current in detector coils (Figure 7.6). A feedback signal processing system is used to 'tune' the forcing to the natural frequency of the cylinder to sustain the oscillations. The natural frequency obtained is measured using a frequency counter.

The natural frequency of the cylinder depends on both the pressure and density of the air inside the cylinder, but shows a complicated non-linear response which has to be obtained empirically by calibration. (The influence of density is also allowed for by measuring the temperature of the sensor.) Precision versions of this instrument are able to measure atmospheric pressure to an accuracy of better than ±0.05 hPa, and with a resolution of 0.01 hPa.

A summary of the different methods used in determining atmospheric pressure is given in Table 7.1.

7.3 Corrections to barometers

For routine use, the pressure measured by a barometer will usually be corrected to sea level. In addition, the effect of flow around a building will cause local pressure differences which can occasionally compromise the measurement from an accurate barometer.

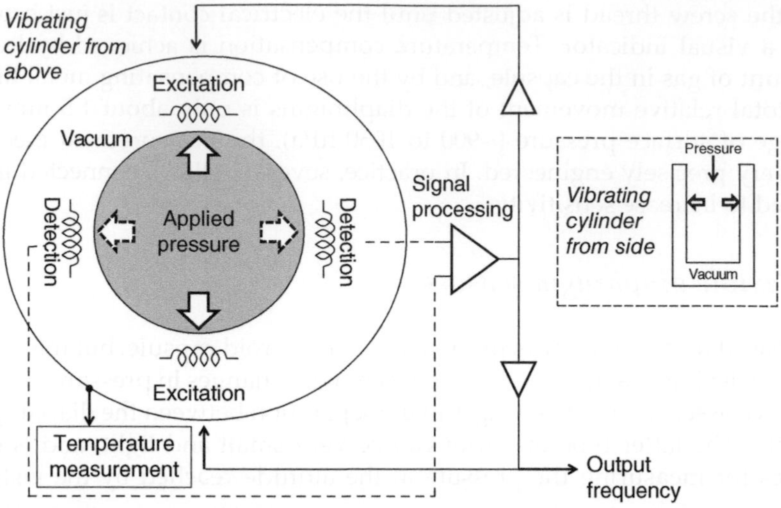

Figure 7.6 Outline schematic of a vibrating cylinder barometer system. Pressure is applied to a thin-walled open-ended central cylinder, which is surrounded by a vacuum. The cylinder is encouraged to oscillate by magnetic excitation applied using coils spaced around the cylinder, with its oscillatory motion detected by other coils. Signal processing circuitry adjusts the frequency of excitation to allow the cylinder to oscillate at its natural frequency. The oscillation is in the hoop mode of the cylinder as viewed from above, alternately across the cylinder in one direction (solid arrows) and then at right angles (dashed arrows).

7.3.1 Sea level correction

Mean sea level is the standard reference height used for pressure measurements at sites below about 1000 m altitude. Hence, for a barometer operated at a site above sea level, a correction due to the additional weight of the air column between sea level

Table 7.1 Sensing methods used in barometry

Pressure sensor	Barometric parameter	Measurement method	Typical accuracy
Mercury column	Column length	Mechanical scale, sometimes with a vernier system	±0.1 hPa
Aneroid capsule	Capsule dimension	Mechanical dial; change in frequency of resonant circuit coupled inductively	±1 hPa
	Precision aneroid device – dimension change of bellows	Micrometer	±0.1 hPa
Hypsometer	Boiling point of a liquid	Temperature	
Vibrating cylinder	Resonant frequency of cylinder	Frequency	±0.05 hPa
Electronic diaphragm	Capacitance between diaphragm and reference plate	Capacitance (via resonant frequency)	±0.5 hPa
	Resistance of strain gauge bonded to diaphragm	Resistance	

and the station height must be added to the barometer reading. If an isothermal atmosphere is assumed (usually a good approximation), the variation in pressure with height z follows an exponential relationship,

$$P_z = P_0 \exp\left(\frac{-z}{H}\right), \tag{7.7}$$

where P_z is the pressure at a height z, P_0 is the sea level pressure and H is the scale height. The scale height is found from $H = R^*T/M_a g$ where T is the assumed isothermal temperature, R^* the universal gas constant, M_a the relative molecular mass of air and g the acceleration due to gravity. A pressure measurement made at a height z can be converted to its equivalent value at sea level by finding

$$P_0 = P_z \exp\left(\frac{z}{H}\right). \tag{7.8}$$

The difficulty in general is deciding on an appropriate isothermal temperature value T to use, and hence there are standard guidelines applied for observing stations. In practice, a value for T can be estimated as the mean temperature between the station and sea level, for example by assuming a typical lapse rate from the station temperature.

7.3.2 *Wind speed corrections*

Air flow around a building housing the barometer causes pressure changes of order $\rho u^2/2$ (see Equation 8.1), which yields appreciable pressure fluctuations compared with the target accuracy of 0.1 hPa, for example 0.4 hPa for $u = 8$ m s^{-1}, although this varies with building geometry and wind direction. For better accuracy under these conditions, barometers can be connected to a static head sampling outside the building, where the flow effect is minimised.

8

Wind Speed and Direction

8.1 Introduction

Although airflow in the atmosphere is a three-dimensional velocity field, it is the horizontal component of its motion which is commonly measured, in part because the vertical component is generally much smaller. Measuring the horizontal air velocity ultimately determines the wind vector $\mathbf{V} = (u, v)$, where u is conventionally the west-to-east (*zonal*) component, and v the south-to-north (*meridional*) component. Often only the magnitude of the vector (i.e. the wind speed, which is usually given the symbol U) is measured, with the wind direction measured entirely independently. An instrument for measuring the wind speed is known as *anemometer*; the mechanical device turned by the wind to indicate the direction from which the wind is blowing is a *wind vane*.

With a fast-response wind sensor, rapid changes of speed and direction can be identified. Fluctuations on all time scales from fractions of a second upwards are present, although changes on timescales longer than about 30 min are not regarded as turbulence. Wind speed measurements spaced along a flow can also show a statistical correlation, when coherent structures or eddies are present.

Properties used to determine wind speed include the flow's kinetic energy (cup and propeller anemometers), pressure (tube, Pitot and pressure plate anemometers), cooling (hot wire or film anemometer) and the effective speed of sound in a moving reference frame (sonic anemometer).

8.2 Types of anemometer

8.2.1 *Pressure plate anemometers*

A simple anemometer is provided by the deflection of a vertically hanging plate oriented into the mean wind direction. The deflection of the plate to the vertical is proportional to the strength of the wind. Although an early experimental method,[i] the non-linear response and the directional response mean it is not widely used. A related

[i] Such a method was proposed for meteorological use by Robert Hooke (1635–1703).

Meteorological Measurements and Instrumentation, First Edition. R. Giles Harrison.
© 2015 John Wiley & Sons, Ltd. Published 2015 by John Wiley & Sons, Ltd.
Companion website: www.wiley.com/go/harrison/meteorologicalinstruments

technique, the deflection of a small weight or lightweight ball by the horizontal wind known as a 'telltale', has however been used as a sensor on the Phoenix Martian lander [71], with the measurement made using a video camera.

8.2.2 *Pressure tube anemometer*

A pressure tube anemometer measures pressure differences associated with a static and moving flow. An early design of this was the pressure tube anemometer of W.H. Dines[ii] in 1892, which measured the small difference in pressure (see Chapter 7 for pressure measurement techniques) from a tube aligned to face the wind by a vane, and a further set of sensing holes exposed to the mean flow. The tube facing the wind determines the total stagnation pressure P_t, and the sensing holes the static pressure P_s.

The wind speed is derived from Bernoulli's principle as

$$\Delta P = P_t - P_s = \frac{1}{2}\rho U^2 , \tag{8.1}$$

where ρ is air density, P_t is the total pressure (from the orifice pointing directly into the air stream) and P_s the static pressure (from an orifice sensing the pressure of the moving air stream).

The pressure difference ΔP is measured using a differential manometer, and is small. For example, Equation 8.1 shows a wind speed of 1 m s^{-1} giving a pressure difference of $\sim$ 0.5 Pa. It is also apparent that the response is non-linear, with least sensitivity at low wind speeds.

A Pitot tube anemometer operates on the same principle, but uses a fixed direction sampling tube pointing towards the incoming flow. For a practical Pitot instrument, an extra 'form constant' should be included in Equation 8.1, although the sensor head is usually designed so that this constant is close to unity. With allowance for the form constant, a well-designed Pitot tube can be regarded as an absolute instrument. Pressure tube anemometers are therefore used for wind tunnel calibrations of other anemometers, but not normally in field measurements because of the directional sensitivity. The Dines pressure tube anemometer has seen long use because of its absolute mode of operation [72].

8.2.3 *Cup anemometers*

A typical cup anemometer[iii] (Figure 8.1) has three conical or hemispherical collecting cups mounted at equal distances from a vertical shaft by equally spaced horizontal supporting arms. The vertical shaft rotates on bearings arranged to cause as little mechanical loading as possible. The asymmetry of the cup arrangement ensures

[ii] Beyond the pressure tube anemometer, William Henry Dines (1855–1928) is also known for his upper air recording work, using the meteograph.

[iii] A four-cup anemometer was developed in 1846 by John Robinson (1792–1882), director of Armagh Observatory.

Figure 8.1 A three-cup anemometer used to measure the horizontal wind speed.

that the anemometer always rotates the same way irrespective of the direction of the incoming wind. The response characteristic of a cup anemometer is close to linear, and can be described by

$$\omega = k(U - U_0), \tag{8.2}$$

where ω is the angular rotation speed, U is the wind speed, U_0 is the starting speed for the anemometer and k a calibration constant. Typically, U_0 is about 2 m s^{-1} for large-cup anemometers used in a climatological station, but may be only 0.5 m s^{-1} for a light anemometer with good bearings as used in micrometeorology.

For small changes of wind speed, the response time constant τ of a cup anemometer is inversely proportional to the wind speed. The product τU is known as the *response length* as it corresponds to a run (or distance) of wind necessary for the instrument to respond, which is fairly independent of the actual wind speed. For a climatological station anemometer, the response length is about 10 m, while for a small lightweight anemometer, it may be only 2.5 m.

The output signal from a cup anemometer may be a voltage proportional to rotation speed (e.g. from a DC generator) or a series of electrical pulses generated by an optical or magnetic switch on the rotating shaft (Figure 8.2). The optical switch is particularly well suited to minimising the rotational load, and hence in obtaining a small starting speed. Both voltage and pulse outputs generally require some smoothing, which limits the time response of the measurement. At slow speeds for pulse output devices, reciprocal counting (timing the interval between pulses, rather than by determining the pulse rate), provides higher resolution of the rotation rate.

Figure 8.2 View of the optical shutter arrangement used in the lightweight cup anemometer of Figure 8.1. As the shaft rotates, a light beam between an optical source and electronic detector is regularly interrupted by the teeth on the shutter, generating a series of digital pulses at a rate which is proportional to the rotation speed.

8.2.4 Propeller anemometer

In contrast to a cup anemometer, a propeller anemometer provides a directional response, and, if the sensing propeller is made of lightweight material, it can be accelerated and decelerated rapidly by the wind. Examples of propeller anemometers are shown in Figure 8.3. These use a 'helicoidal' design of lightweight propeller, mounted on a shaft to drive a small DC generator [73]. If the anemometer shaft is inclined at a small angle θ to a steady air flow of speed U, the rotation rate of the shaft is expected to be given by

$$\omega = k(U - U_0)\cos\theta\,. \tag{8.3}$$

The starting speed is much lower than for most cup anemometers (typically 0.2 m s^{-1}), and response lengths correspondingly smaller (typically 1 m). Consequently, this type of anemometer is suitable for measuring turbulent fluctuations of air speed along one direction. Unfortunately, Equation 8.3 is only approximated for relatively small angles (typically less than 30°), due to the effect of one blade interrupting the flow to another blade, known as blade sheltering. Consequently, if a propeller anemometer is only required to measure wind speed, then it is usually kept oriented along the direction of the wind with the aid of a wind vane.

A common application of the propeller anemometer is to measure turbulent fluctuations in the vertical component of the wind (often given the symbol w). If simultaneous measurements of temperature using a fast response thermometer are

Figure 8.3 An array of three orthogonal propeller anemometers providing a rapid response to wind fluctuations, from which the wind direction can also be found. (A fast response thermometer is also mounted in the middle of the anemometer array.)

also made, their combination provides an estimate of the vertical transfer of heat (see Section 12.1.2). For this application, the anemometer is mounted with its axis vertical.

8.2.5 Hot sensor anemometer

A hot sensor anemometer determines the wind speed by the amount of cooling generated from a heated sensor. The sensor may be in the form of a fine wire (a hot wire anemometer) or the bead of a thermistor (a hot bead anemometer), to which a heating current is supplied. Two operating modes are possible; in one case by measuring the

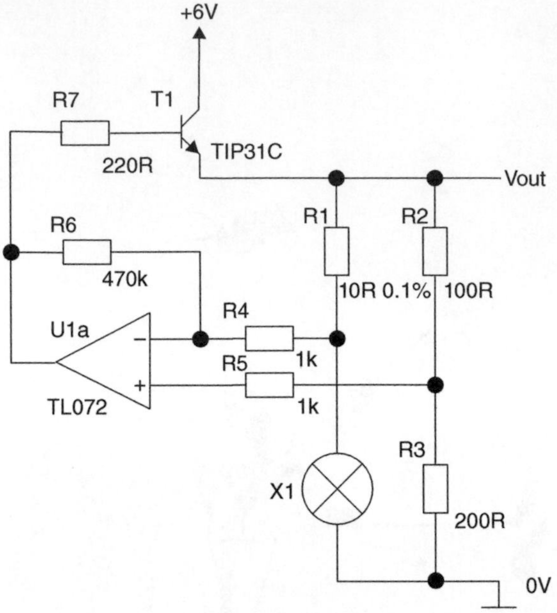

Figure 8.4 Principle of the electronic feedback system used to control a constant temperature hot wire anemometer. The hot wire anemometer sensing element (X1) has a filament lamp with the glass removed, a resistance which varies with temperature, and this forms a potential divider with R1. The mid-point voltage of this potential divider varies with the element's temperature, which varies in turn with the air flow. This mid-point voltage is compared (by differential operational amplifier U1a), with the reference voltage from a potential divider (R2 and R3), which is insensitive to the air flow. The difference voltage is amplified and used to bring the two mid-point voltages into agreement, by controlling the current supplied to both dividers by the transistor T1. This feedback stabilises the hot wire sensor temperature, with the excitation voltage of the potential dividers (V_{out}) providing the signal to be measured.

power needed to keep the sensor at constant temperature, or, in the other, by measuring the sensor's temperature with a constant supply of power.

In the case of a heated fine wire suspended in an airflow, the wire also operates as a thermometer, in that its resistance is proportional to its temperature. It can be kept at a constant temperature electronically by a feedback circuit, which varies the heating current to maintain a constant resistance (Figure 8.4). For this situation, the current supplied to the wire to keep its temperature constant will be determined only by the wire's cooling rate, and therefore provides the quantity to be measured. A general empirical relation between heating current and air speed, known as King's law, relates the voltage drop across the wire to wind speed by

$$V^2 = A + B\sqrt{U}\,. \tag{8.4}$$

Such a relationship is obviously non-linear, and shows that the sensitivity increases as a power law with decreasing wind speed. (The response of a practical sensor measured in a wind tunnel is shown in Figure 8.5.) Because such a sensor can be

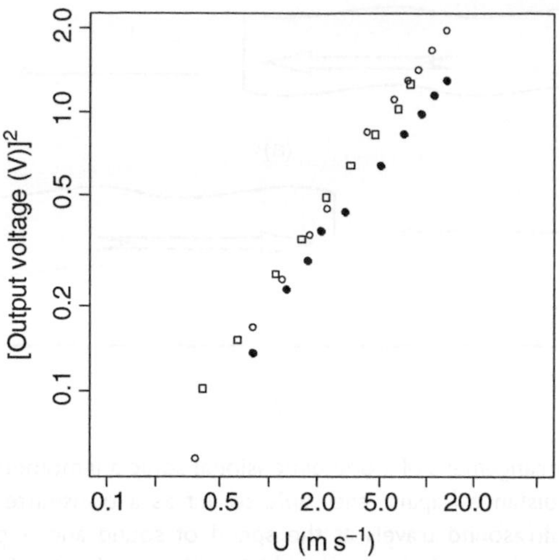

Figure 8.5 Measured wind speed response of a hot wire sensor, using the constant temperature control principle of Figure 8.4. Three similar sensors were used, each identified by different plot symbols, which demonstrates the need for individual sensor characterisation.

constructed to be small, the response can be rapid, so hot wire (or hot bead, or even a filament bulb with the glass envelope carefully removed [74]) anemometers are normally used to measure turbulence or details of flow structure over objects mounted in a wind tunnel. They are, however, sufficiently delicate that they are not well suited to field measurements, although the hot bead devices can be mounted in a protective head to increase their robustness. The response of the wire or bead is also directional, and therefore consistent alignment with respect to the flow is important if the directional response is not to contribute to the measurement.

8.2.6 *Sonic anemometer*

The time of flight of a pulse of sound propagating in air provides a further method for sensing the wind speed. Ultrasound[iv] can be readily generated and detected electronically, and is used for such measurements. In an ultrasonic anemometer, the flight time required for pulses of ultrasound to travel forwards and backwards between two fixed transducers A and B is measured (see Figure 8.6). The separation of the sensors is usually 10 to 20 cm, and as these measurements are electronic, they can be repeated rapidly, at typically 5 to 100 Hz, giving good accuracy and time response.

If the flight time is t_1 from A to B

$$t_1 = \frac{l}{(c_s + v)}, \tag{8.5}$$

[iv] Ultrasound is acoustic energy which has frequencies above the usual range of human hearing, such as sound in the frequency range ~40 kHz to 100 kHz used for sonic anemometry.

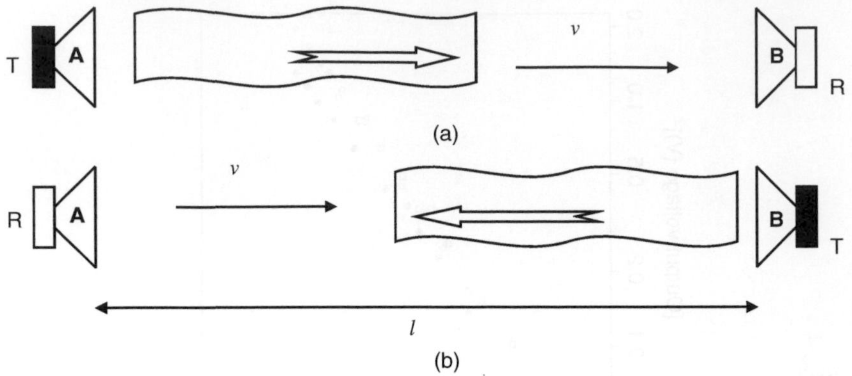

Figure 8.6 Conceptual arrangement of a one-dimensional sonic anemometer, with reversible transducers A and B spaced a distance *l* apart, each able to act as a transmitter (T) or receiver (R) of ultrasound. The pulse of ultrasound travels at the speed of sound and is propagated forwards (a) and backwards (b), encountering a wind speed *v* which adds or subtracts from the speed of sound accordingly.

and t_2 from B to A,

$$t_2 = \frac{l}{(c_s - v)}, \tag{8.6}$$

where v is the component of wind travelling from A to B and c_s is the speed of sound. Eliminating c_s gives

$$v = \frac{l}{2}\left\{\frac{1}{t_1} - \frac{1}{t_2}\right\}. \tag{8.7}$$

Alternatively, c_s can be found, which is strongly temperature-dependent. This variation is given by

$$c_s = \sqrt{\frac{\gamma R^* T}{M_r}}, \tag{8.8}$$

where T is the air temperature, R^* the universal gas constant, M_r the relative molecular mass of air and γ the ratio of the specific heat capacities of air at constant pressure and volume, c_p/c_v. Inserting typical values for dry air ($\gamma = 1.4$, $M_r = 0.029$ kg mol^{-1} and $R^* = 8.31$ JK^{-1} mol^{-1}) gives c_s at room temperature ($T = 298$ K) as ~346 m s^{-1}. If γ and M_r are assumed to be constant (i.e. considering the air to be dry), the *sonic temperature* T_s can also be defined from Equation 8.8, such that

$$c_s^2 = \frac{\gamma R^*}{M_r} T_s \approx 403 T_s. \tag{8.9}$$

This relationship forms the basis for sonic thermometry, where a measure of the air temperature is found from the speed of sound. However, a complication arises if the air is moist, as γ and M_r then need to be modified [75] to allow for the partial contribution of water vapour, giving

$$c_s^2 = 403T\left[1 + 0.32\frac{e}{p}\right], \tag{8.10}$$

where e is the water vapour pressure and p the absolute pressure. This shows the relationship between the sonic temperature and air temperature, as

$$c_s^2 = 403T\left[1 + 0.32\frac{e}{p}\right]. \tag{8.11}$$

For $e \ll p$, the sonic temperature becomes a good approximation to the air temperature.[v] Consequently, the humidity correction effect is usually neglected, particularly when using sonic thermometry to determine air temperature fluctuations. The difference between the air temperature and sonic temperature, if available very accurately, does, however, yield an acoustic method for measuring relative humidity [76].

Because of its rapid response, a sonic anemometer is well suited to the field measurement of turbulent fluctuations of wind velocity, and, as Figure 8.7 shows, in several directions simultaneously. After applying Equation 8.7, the three-dimensional velocity vector can be calculated with some data processing and trigonometric identities. The only significant disadvantages of this instrument are cost and management of the large volumes of data which can be quickly generated.

8.3 Wind direction

Variations in wind direction can provide sensitive detections of changes in conditions, such as those associated with the passage of fronts, sea breezes or even solar eclipses [77]. These are complementary to other changes occurring which may be obscured by more substantial variations due to diurnal cycles.

Wind direction is conventionally defined as the bearing of a point from which the air is blowing, reckoned clockwise from north. This meteorological convention means that the wind direction is considered to be in the opposite direction to that of the vector describing the horizontal air velocity. Thus, for example, a 'south-westerly' wind would describe a wind direction of 225°, but its vector direction would be 45°. A wind direction of 0° is sometimes used to signify calm conditions in databases as a null variable, as no wind direction is then defined.[vi] It is important to identify when calm conditions affect wind vanes, as then the wind direction indicated merely refers to the last wind direction before the conditions became calm. In light winds inconsistent starting speeds of vane and cup anemometers may also cause uncertainties.

---

[v] The sonic temperature is similar to the *virtual temperature*, which is the temperature at which a sample of dry air would have the same density as a sample of moist air, for no associated change in pressure.

[vi] As discussed in Chapter 4, it is preferable to avoid assigning a real number to a data value which cannot be obtained, and instead to mark it as unavailable, to prevent erroneous results occurring in automated processing.

Figure 8.7 Ultrasonic anemometers mounted on masts. In the foreground is a two-dimensional device, and in the background a three-dimensional device. For the three-dimensional anemometer, the three pairs of transmitting and receiving transducers are towards the centre, with an outer support frame arranged to minimise disturbance to the flow.

8.3.1 Wind vanes

For synoptic and climatological applications, wind direction is usually measured simply using a wind vane. Except for decorative versions (see, e.g. Figure 8.8), a wind vane usually consists of a flat vertical plate attached to a horizontal arm able to rotate on a vertical shaft. Such a device operates on the simple principle that the dynamic wind pressure P ($= 0.5\rho U^2$) causes a static force on a plate if the plate is not exactly aligned with the wind direction. The direction of the arm therefore becomes aligned with the wind direction.

The response of a wind vane to changes in wind direction is limited by the need to damp the motion in order to avoid overshoot. It is theoretically possible to arrive at a critical damping factor, such that the vane responds quickly without overshooting. Practical designs allow slight overshoot with rapid damping of oscillations about the mean wind direction.

The direction of the arm of a mechanical wind vane can be determined by several methods:

(i) from a mechanical switch with multiple contacts distributed regularly around the shaft;
(ii) from an optical encoder giving a digital representation of measured angle; and
(iii) from a precision potentiometer attached to the shaft, able to rotate continuously.

Figure 8.8 Ornate weather vanes receive varying amounts of maintenance, and are unlikely to have low friction bearings. A combination of several devices provides more confidence in the direction indicated.

The principle of a potentiometer wind vane is shown in Figure 8.9. A potentiometer provides a fixed resistance between the two ends of a track of resistive material, and a movable contact (or wiper), which can be positioned anywhere between one end of the track and the other. If the two ends of the resistive track are connected across a fixed voltage supply, as the wiper moves from one end to the other, the voltage on the wiper will vary smoothly across the supply range from zero to its maximum value. For a track constructed in a circular form, this will provide an output voltage proportional to the amount of rotation, except for a small dead zone where the resistance track begins and ends.

Although this will provide an accurate measurement of the rotation, such an approach also requires that the orientation of the vane is already known or has been

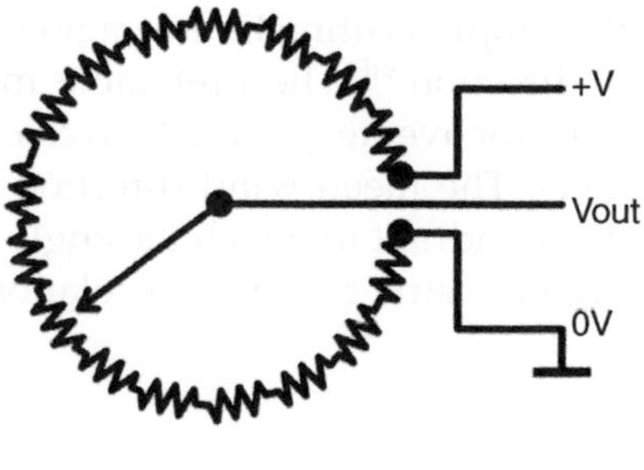

Figure 8.9 Principle of a potentiometer wind vane. Viewed from above (left hand side of diagram), the potentiometer wiper is free to rotate continuously for almost all the 360° of the resistance track, with a small dead zone necessary between the beginning and end of the track. When powered from a regulated supply, the voltage V_{out} at the wiper is proportional to the angular rotation.

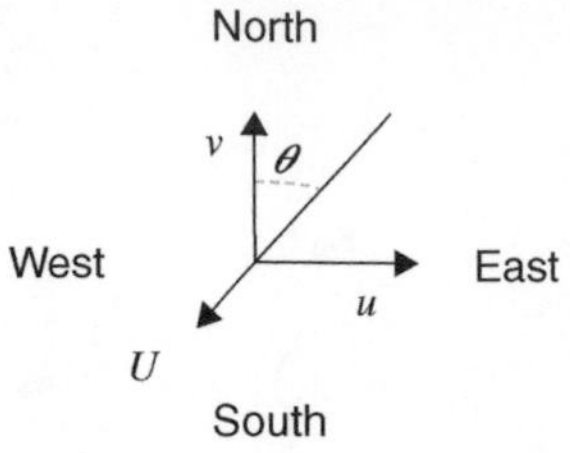

Figure 8.10 View from above of a horizontal wind of magnitude U blowing from the north-east, with a wind direction θ reckoned clockwise from north. The wind components in the west-east and south-north directions are u and v respectively.

correctly installed to a standard configuration. For some circumstances where this cannot be assumed, such as on ocean buoys, a magnet may be used to orientate the direction measuring instrumentation, giving a self-referencing wind vane.

8.3.2 *Horizontal wind components*

An array of two or three directional instruments (Figure 8.3) or a multi-direction instrument (Figure 8.7) allows measurements of both wind speed and direction. Figure 8.10 shows a wind of speed U at a bearing θ from north, which can equivalently be expressed in terms of wind components u and v (in the west-east and the south-north directions respectively) by

$$u = -U\sin\theta\,, \tag{8.12}$$

and

$$v = -U\cos\theta\,. \tag{8.13}$$

Resolving the wind speed and direction into u and v or other wind speed components is essential if the wind direction needs to be averaged. This is because wind direction is a circular measure, and its value repeats as the wind direction passes through north. Consequently simple arithmetic averaging of wind direction may generate entirely the wrong direction.[vii] The preferable method is to first derive wind speed components u and v for averaging, and then calculate the wind direction from the averaged components. The mean wind direction is given by $\tan^{-1}(v/u)$, although careful account of the quadrant in which the angle is returned is needed[viii] especially as some inverse trigonometric function calculations only return angles in the range ±90°.

[vii] Consider, for example, averaging two wind directions when the wind is blowing from around north, of 10° and 350°. The arithmetic average of these two values is 180°, which represents a southerly wind, whereas the actual mean wind direction is 360°, i.e. northerly.

[viii] Some computer languages offer an inverse tangent function *atan2* which returns angles in the range ±180°, but note there are variations in calling conventions for the order of the arguments between different languages.

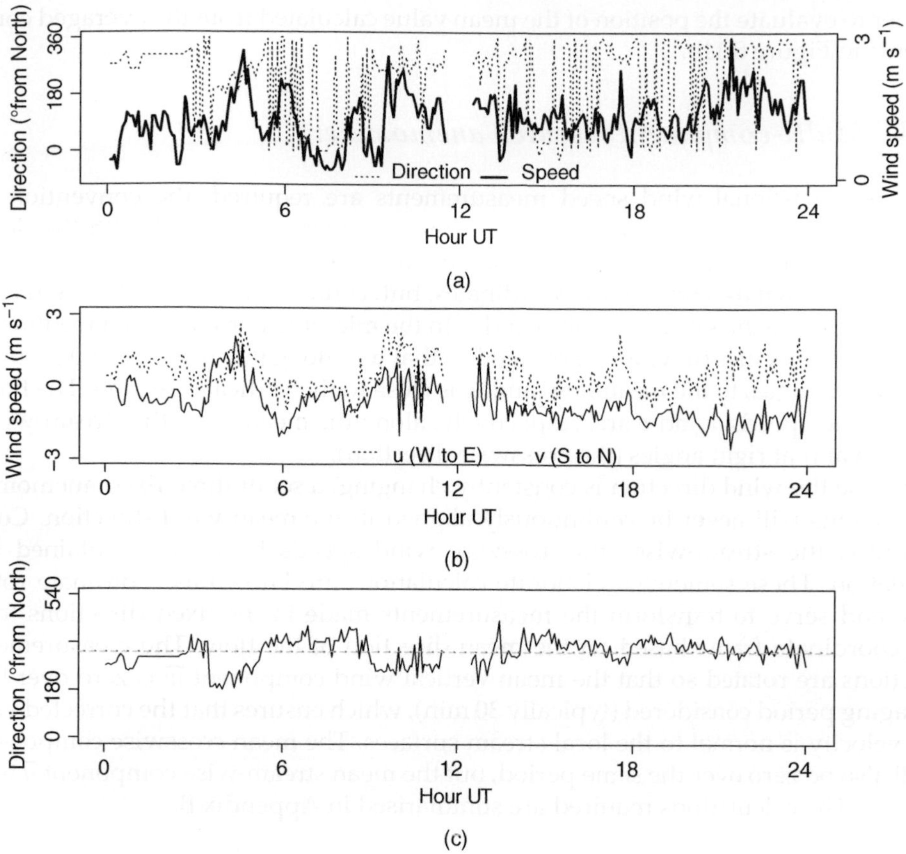

Figure 8.11 (a) Scalar wind speed and direction measured at 10 m at Reading, on a day with flow mostly around northerly (10 January 2013). (b) shows the calculated south-north and west-east components. (c) shows the wind direction recalculated from the wind components in (b), but offset to allow the wind direction in the first (NE) quadrant to be reported in the range 360° to 540°, with the calculated mean value also marked (dashed line).

Figure 8.11 illustrates how the mean wind direction can be obtained on a day when the wind is northerly. Figure 8.11a shows the time series of the wind speed and direction, obtained using a wind vane and cup anemometer at 10 m. Clearly the wind direction is fluctuating around northerly, and although substantial changes are apparent in wind direction, generating lines between the top and bottom of the plot, the actual fluctuations in wind direction are small in terms of absolute change in angle. Figure 8.11b shows the two horizontal wind speed components calculated from the wind speed and direction in Figure 8.11a which are both single values and hence can be separately averaged and then the mean wind direction found from trigonometry. An alternative approach sometimes adopted is to extend the range of wind directions before the direction returns to zero. Figure 8.11c shows the wind direction in this way, to prevent the wind direction values beyond 360° resetting to 1°, instead allowing the wind direction to carry on increasing. This has the effect of removing the appearance of substantial wind direction fluctuations from the plot. It also makes it easier to evaluate the position of the mean value calculated from the averaged components in Figure 8.11b.

8.3.3 *Multi-component research anemometers*

If three-dimensional wind speed measurements are required, the convention in micrometeorology is to adopt wind components which are aligned with the flow, rather than the fixed compass directions of the usual meteorological convention. These are known as stream-wise coordinates, but, confusingly, the related literature also tends to use the same symbols u and v. In the micrometeorological convention, a set of three velocities (**u**, **v**, **w**) is arranged so that **u** is along the mean wind direction, **v** is at right angles to the mean flow and **w** is close to the vertical direction. The horizontal wind speeds u and v are, respectively, along the mean flow (the stream-wise) direction and at right angles (the crosswise direction).

Because the wind direction is constantly changing, a set of three fixed anemometer positions will never be continuously aligned in the mean wind direction. Consequently, the stream-wise and crosswise wind speeds have to be obtained by calculation. These somewhat elaborate calculations are known as coordinate rotations, and serve to transform the measurements made in the fixed directions to a new coordinate frame based on the mean direction of the flow. The measurement directions are rotated so that the mean vertical wind component $\overline{w}$ is zero over the averaging period considered (typically 30 min), which ensures that the corrected vertical velocity is normal to the local stream surfaces. The mean crosswise component $\overline{v}$ will also be zero over the same period, but the mean stream-wise component $\overline{u}$ will be finite. The calculations required are summarised in Appendix B.

8.4 Anemometer exposure

For synoptic meteorology, wind speed and direction is conventionally measured at 10 m above the surface with the wind unobstructed in all directions. These can be achieved directly by using instruments mounted at that height (Figure 8.12) or by applying a correction to measurements made at the greatest height available. Use of such a correction is possible because, at an unobstructed site, the wind speeds measured at different heights vary with a known vertical profile (see also Figure 12.6).

8.4.1 *Anemometer deficiencies*

A key requirement in the operation of mechanical anemometers and wind wanes is that their bearings maintain low friction for minimal loading. Snow accumulations provide one source of obstruction which can affect cup anemometers, leading to a change in their calibration. At high wind speeds, the snow may fall out of the cups, but at low speeds the combination of low temperatures and clogged cups may even lead to the sensor freezing. Heated versions exist which can circumvent this problem but power is then required which may not be available at remote sites.

Figure 8.13 shows a comparison between the horizontal wind speed determined by cup and sonic anemometers at 10 m during snow. The wind speed measured by the

Figure 8.12 Standard meteorological mast at Camborne Met Office, Kehelland, Cornwall, carrying anemometers and wind vanes.

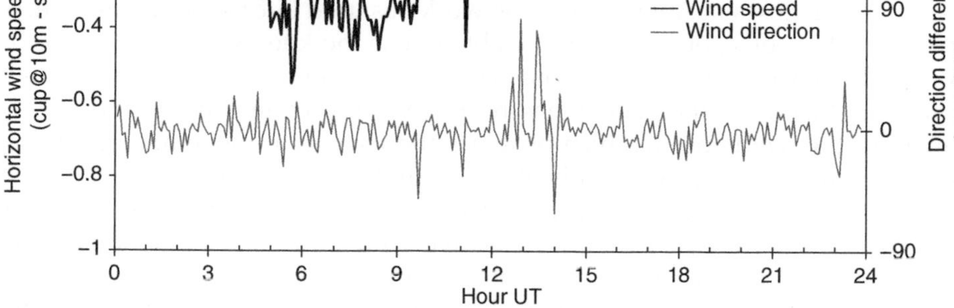

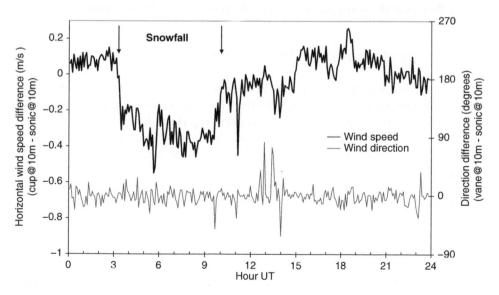

Figure 8.13 The effect of heavy snowfall (8 February 2007) on the difference between the 5-minute mean wind speed at 10 m measured by a cup anemometer and three-component sonic anemometer at Reading. The difference in wind direction determined by a wind vane and the same sonic anemometer is also shown (the mean daily wind speed was 2.5 m s^{-1}, standard deviation 1.2 m s^{-1}).

Figure 8.14 A Rokkaku kite (flat area 2.3 m^2) with a tension-measuring sensor at the base of the tether, flown close to a sonic anemometer operating at 10 m.

cup is reduced during the snowfall. There is no simultaneous anomaly in a comparison between the wind direction determined from the same sonic anemometer data and the wind direction from a wind vane. This indicates that it is more likely to be the cup rather than the sonic anemometer which is affected by the snowfall.

8.5 Wind speed from kite tether tension

Wind speed measurements immediately above the surface generally use meteorological towers or adjustable masts,[ix] but tethered kites can also be used for wind speed measurements when conditions are changing and rapid deployment is needed, such as for monitoring flows in Antarctica. Some work has used kites as aerial instrument platforms or 'skyhooks', but it is also possible to use the kite itself as the sensor by measuring the tension in the tether line. This operates on the same principle as a wind vane, in that, for a flat plate, the wind force is approximately proportional to the square of the wind speed, which provides the tension in the tether. The tether tension can be monitored conveniently at the ground anchoring point.

A strain gauge can be used to measure the tension, but an important aspect of such measurements, as the tension variation is relatively small for a kite of modest dimensions, is to ensure that the instrumentation is not strongly influenced by temperature. In the kite wind-monitoring system shown in Figure 8.14, a measuring anchor ring was used with strain gauges mounted around it to monitor the distortion under tension [78].

[ix] Hollow telescopic metal masts, which can be pumped up to achieve an extension to the height required, deflated afterwards for convenient transport, provide robust and transportable options for field work.

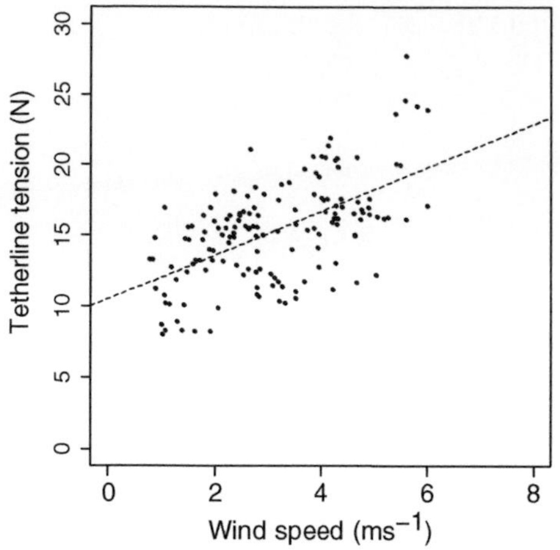

Figure 8.15 Response of the kite tether line tension to wind speed, for instantaneous (1-second sampling) measurements of tension and mean wind speed determined by a nearby sonic anemometer at 10 m. (An extrapolation to zero wind speed is marked to indicate the zero offset in tension.)

Figure 8.15 shows the response obtained in the tether line tension, during which the mean wind speed was determined by a nearby sonic anemometer. The instantaneous tension in the tether line shows, approximately, a linear response to the instantaneous wind speed measurements. Some scatter arises from the comparison between the point measurement and area average obtained by the kite.

9

Radiation

9.1 Introduction

Radiation measurements serve to determine the rate at which energy arrives at a surface of known area. Radiation can variously be expressed as *radiant flux* (the radiation energy received, emitted or transmitted per unit time), *radiance* (radiation falling within a given solid angle), *radiant flux density* (radiative flux per unit area), *irradiance* (radiant flux density incident on a surface) or *emittance* (radiant flux density emitted from a surface). A *radiometer* is any instrument which measures radiation.

The range of wavelengths at which a body emits radiation depends on its temperature, as described by Planck's law (see Figure 9.1).

Calculations from Planck's law show that the wavelength at which maximum radiation is emitted is inversely proportional to temperature.[i] The natural radiation field at the earth's surface can therefore conveniently be divided into two distinct broadband spectral regions, the 'shortwave' (solar) radiation (approximate wavelengths 0.3 to 3 μm) and 'longwave' (terrestrial) radiation (wavelengths 3 to 100 μm).

The amount of radiation reaching the base of the atmosphere depends on complex multiple interactions (such as scattering and absorption) by clouds, aerosol and atmospheric gases, and the emission of longwave radiation (Figure 9.2).

Shortwave measurements made at the surface can determine

1. the direct solar beam irradiance S_b, which is the radiation received directly from the sun, usually measured normal to the beam;
2. the diffuse solar irradiance S_d, which is the scattered solar radiation;
3. the global solar irradiance, S_g, which is the total (scattered and direct) shortwave radiation falling on a horizontal surface; and
4. the upward solar irradiance S_u reflected from the horizontal surface (usually given by αS_g, where α is the reflection coefficient).

[i] This result is also known as Wien's law.

Meteorological Measurements and Instrumentation, First Edition. R. Giles Harrison.
© 2015 John Wiley & Sons, Ltd. Published 2015 by John Wiley & Sons, Ltd.
Companion website: www.wiley.com/go/harrison/meteorologicalinstruments

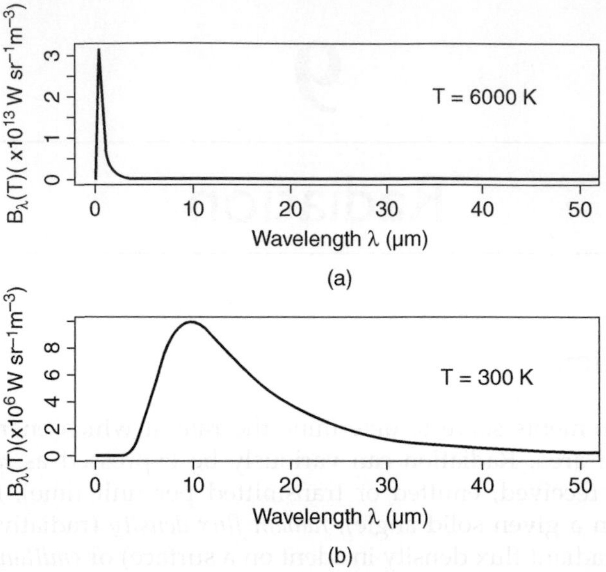

Figure 9.1 Variation of spectral radiance with wavelength calculated from Planck's law, for bodies radiating at two different temperatures typical (a) of the sun's surface and (b) the terrestrial environment.

As shown in Figure 9.2, upwelling and downwelling longwave radiation (L_u and L_d respectively) are also present in the atmosphere. Considered together, the incoming and outgoing longwave and shortwave radiations determine the net radiation R_n arriving at the surface, as

$$R_n = (S_g - S_u) + (L_d - L_u). \tag{9.1}$$

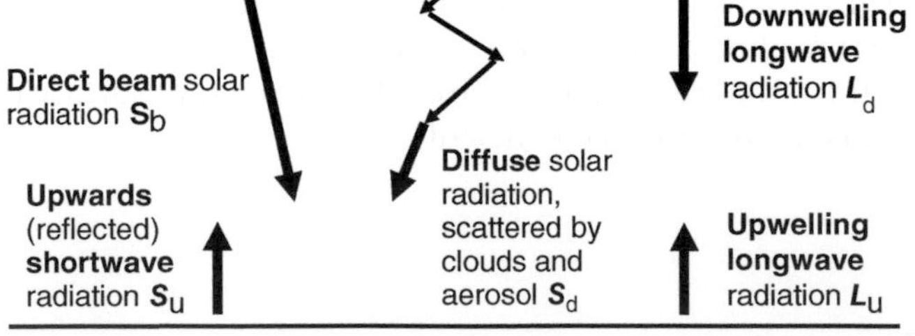

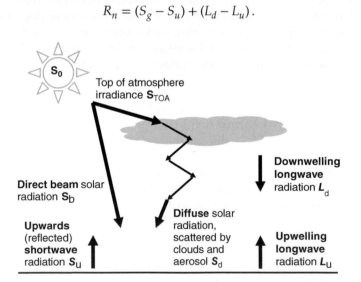

Figure 9.2 The atmospheric radiative environment to be measured at the surface. This consists of longwave radiation emitted upwards from the surface (L_u), downwards from clouds and gases (L_d), and solar radiation entering the top of the atmosphere (horizontal irradiance S_{TOA}) attenuated as it passes in a beam downwards (S_b) or scattered to become diffuse (S_d). Some of the solar radiation received is reflected upwards (S_u). The sun's output is known as the total solar irradiance (S_0).

Radiometers detect radiation by

1. conversion of the electromagnetic energy received into thermal energy by an absorbing material covering the surface of a thermopile; and
2. the response of a semiconductor to photons arriving at the surface of the material, generating an electrical current or voltage.

In the first case, the temperature difference between the two sides of the thermopile measures the difference in radiant energy absorbed by its two surfaces. The Moll thermopile sensor – a series connection of constantan-manganin® thermocouples – provides an electrical output across a wide range of wavelengths of incident radiation, that is it has a *broadband* response. In contrast, the output from semiconductor sensors is wavelength-dependent, and may also be temperature-dependent. Consequently, semiconductor detectors are less commonly used for atmospheric radiation measurements.

For thermopile sensor instruments, the typical output signal is 5–10 mV during sunny UK summer conditions. Considerable voltage amplification (×100 to ×500) is therefore necessary for such low-level signals to be recorded by high-level (e.g. ±5 V full scale) data logging systems. It is important that such amplification is itself thermally stable, and does not contribute additional noise to the measurement (see Section 3.3.2 for description of an amplifier suitable for use with thermopile radiometers to allow their output to be recorded by a data acquisition system).

Table 9.1 summarises instruments used for solar and terrestrial radiation measurements which will be discussed further in Section 9.3 and Section 9.7 respectively. The difference in the range of wavelengths measured depends largely on the filtering action of the protective cover used with the sensing surface, such as the glass used for the shortwave instruments, which does not transmit longwave radiation. Firstly,

Table 9.1 Summary of instruments used for radiation measurements

Wavelength range	Instrument	Purpose
Shortwave	Pyrheliometer	Measurement of direct solar radiation, normal to the solar beam
	Pyranometer	Measurement of global solar radiation over a hemispheric field of view
	Photometer	Measures light in the visible part of the spectrum only, either as energy or number of quanta
	Albedometer	Measures shortwave radiation received by and reflected from a surface
Longwave and shortwave	Pyrradiometer	Measurement of the net radiative flux of shortwave and longwave radiation
Longwave	Pyrgeometer	Measurement of terrestrial radiation, usually over a hemisphere, upward or downward
	Infrared thermometer (or pyrometer)	A radiometer measuring longwave radiation emitted from an object, from which the object's temperature is derived
Any	Spectro-radiometer	Measurement of the radiative energy in a specific part of the spectrum at a given spectral resolution

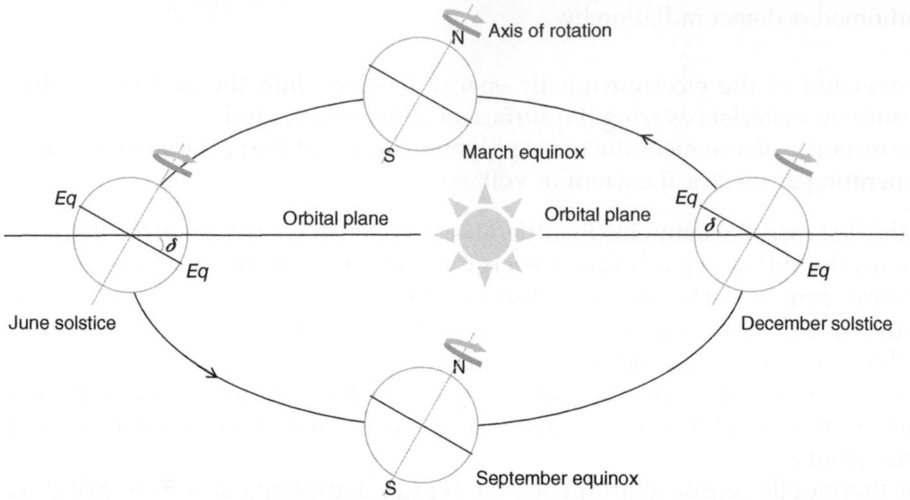

Figure 9.3 Geometry of the sun-earth system, for the earth's position at the solstices and equinoxes. The declination δ is the angle between the orbital plane and the equator (marked as Eq–Eq, perpendicular to the north–south NS axis), as viewed from the sun. The declination is considered positive at the June solstice and negative at the December solstice.

however, the calculation of solar radiation expected from solar geometry calculations will be considered, as this determines important parameters useful for solar radiation measurements, such as the day length, the top of atmosphere solar radiation and correction of diffuse solar radiation measurements.

9.2 Solar geometry

The solar radiation entering the atmosphere at a given time and place (as defined by its longitude and latitude, ϕ) is determined by the earth's rotation and orbital motion. Calculation of the diurnal variation in solar radiation at the top of the atmosphere requires the combination of a seasonal effect associated with the earth's orbital variation, and the diurnal variation arising from the position of the sun in the sky.

9.2.1 *Orbital variations*

The earth's orbit is elliptical.[ii] Seasonal variations in solar radiation arise at earth because the earth's axis is tilted with respect to the orbital (ecliptic) plane (see Figure 9.3). As the sun's rays travel to earth parallel to the ecliptic, the angle between the equator and ecliptic determines the proportion of the sun's rays reaching earth. This angle is known as the *declination* δ, and varies from +23.45° (at the June solstice) to −23.45° (at the December solstice).

[ii] The earth's orbit has a slight eccentricity. Earth is currently slightly nearer the sun in the northern hemisphere winter, at *perihelion*, January 4, than in the northern hemisphere summer, at *aphelion*, July 5. This contributes a small part of the variation in solar energy at the surface.

Accurate calculation of the declination on any given day of the year d requires elaborate calculation using detailed orbital parameters, but it can be approximated as

$$\delta = 23.45^\circ \sin\left[\frac{(d-81)}{365}360^\circ\right]. \qquad (9.2)$$

9.2.2 *Diurnal variation*

Considering the daily variation at a specific location, the angle of the sun in the sky to the local vertical is known as the *solar zenith angle* Z, or alternatively, in terms of its complement, the *solar elevation*. In the northern hemisphere, the *solar azimuth* A is the angle between the sun and true south projected onto the local horizontal plane, negative before noon and positive after. At any point on the earth's surface, the angle between the direction of the sun and local vertical depends only on the latitude and the time. Time can therefore be measured as the *hour angle* (h), which is the fraction of one rotation through which the earth has turned since the apparent local solar noon (when $h = 0$). As the earth turns 360° (or 2π) in 24 hours, the hour angle h in degrees is therefore

$$h = \frac{(t-t_0)}{24}360^\circ, \qquad (9.3)$$

where t is the time and t_0 is the time of solar noon, each in hours.

9.2.3 *Solar time corrections*

The *apparent local solar time* is the time shown by a sundial at a particular location. It does not exactly equal the time as shown by an accurate, standardised clock, as is apparent from a simultaneous comparison of the two (see Figure 9.4).

The solar time can be calculated from the local standardised 'clock' time (e.g. UT or GMT), firstly by subtracting 4 min for each degree west from the Greenwich meridian, and secondly by applying an additional correction known as the *equation of time*. This correction [79] varies with time of year because of the elliptic orbit (with small effects from the eccentricity) and variations in tilt (obliquity) of the earth's rotation axis, reaching its greatest value (±15 min) in February and November (Figure 9.5a). The equation of time is also considered in terms of solar declination[iii] in Figure 9.5b.

The equation of time gives the difference between standard time and solar time, and essentially provides a correction for sundials so that the time indicated by the sundial can be read as standardised time. Although a complicated variation, some sundials actually incorporate this correction directly (see Figure 9.6).

[iii] This is known as an *analemma*, a curve which represents the change in angular offset of a celestial object as viewed from another celestial object.

Figure 9.4 Comparison of standard time (as shown on the face of Westminster Clock tower) with sundial time (as shown on a sundial of St Mary's Westminster), on 4 August 2012, at a location close to the Greenwich meridian.

9.2.4 *Day length calculation*

The day length defines the duration of solar heating at a particular location. It can be found from the basic equation for the zenith angle Z (the angle between the solar beam and the local vertical), which, in terms of the hour angle h, latitude ϕ and declination δ is

$$\cos Z = \sin\phi \sin\delta + \cos\phi \cos\delta \cos h\,. \tag{9.4}$$

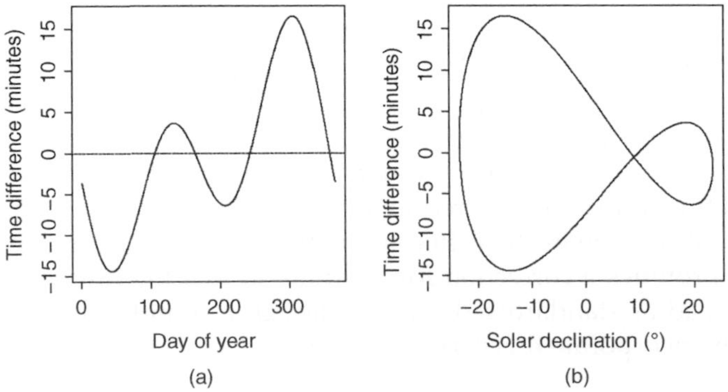

Figure 9.5 (a) The 'equation of time', which provides a correction between solar time and standard time as the year progresses. (b) Equation of time plotted in terms of solar declination.

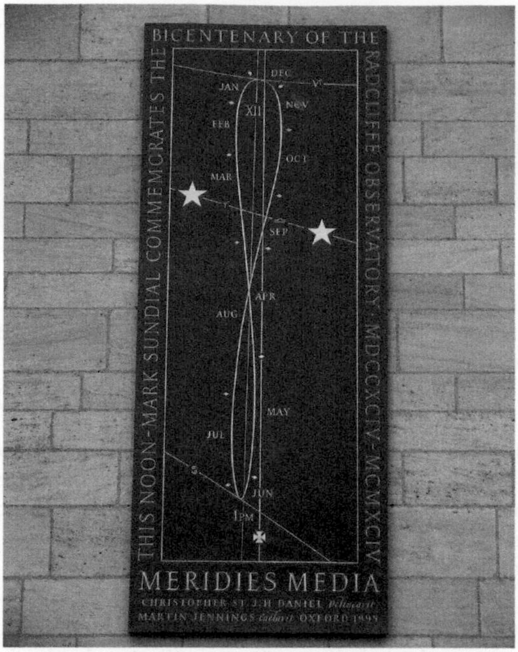

Figure 9.6 Sundial in which the equation of time correction has been incorporated to allow the displayed solar time to be read as standard time (Green Templeton College, Oxford).

The day length (or duration of daylight when the sun is above the horizon) can be calculated from Equation 9.4, by assuming, at sunset, $Z = 90°$ (i.e. $\cos Z = 0$), so the hour angle h_0 between noon and sunset is

$$\cos h_0 = -\frac{\sin\phi \sin\delta}{\cos\phi\cos\delta} = -\tan\phi\tan\delta\,. \tag{9.5}$$

As the day is symmetric about its solar noon, the total day length is $2h_0$. Using Equation 9.3 for h, the day length l in hours can be given in terms of latitude and declination as[iv]

$$l = 2\left(\frac{24}{360°}\right)\cos^{-1}(-\tan\phi\tan\delta)\,. \tag{9.6}$$

9.2.5 *Irradiance calculation*

The irradiance of a surface normal to the sun's beam at the mean distance of earth from the sun (1.50×10^{11} m) as measured by satellite instruments is the *Total Solar*

[iv] This neglects solar refraction effects, and the fact that sunset and sunrise occur when the upper rim of the solar disc is just visible. Some radiation also remains during twilight; *civil*, *nautical* and *astronomical* twilights are defined when the sun is 6°, 12° and 18° below the horizon respectively.

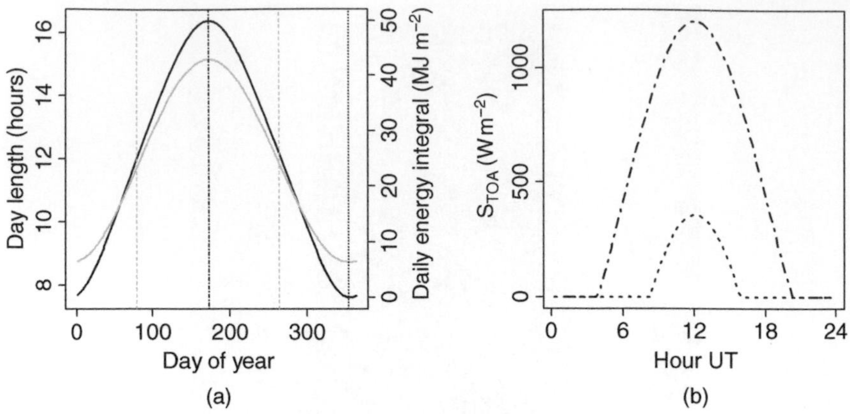

Figure 9.7 Solar radiation calculations made for Reading (latitude 51.442°N, longitude 0.938°W). (a) Day length (black line) and integrated daily energy at the top of atmosphere (grey line) across the year, with grey vertical lines marking the equinoxes, dot-dashed black line the summer solstice and dashed black line the winter solstice. (b) Daily top of atmosphere irradiance on a horizontal surface, for the summer (dot-dashed black line) and winter (dashed black line) solstice.

Irradiance (TSI), which is approximately[v] 1365 W m^{-2}. Hence, for a horizontal surface at the top of the atmosphere, the irradiance S_{TOA} is the total solar irradiance projected onto a surface normal to the solar beam, found from the zenith angle as

$$S_{TOA} = S_0 \left(\frac{d'}{d}\right)^2 \cos Z, \tag{9.7}$$

where d' and d are the actual and mean sun-earth distances respectively, and S_0 is the total solar irradiance.

The daily total amount of radiation per unit area (i.e. a quantity with units of J m^{-2}) found by substituting for cos Z from Equation 9.4 and integrating is

$$I_{TOA} = \frac{86400}{\pi} S_0 \left(\frac{d'}{d}\right)^2 (h_0 \sin\phi \sin\delta + \cos\phi \cos\delta \sin h_0), \tag{9.8}$$

where h_0 is in radians and the $(d'/d)^2$ term is small (± 3%). Using these results, Figure 9.7 shows (a) the calculated day length and daily integrated top of atmosphere energy and (b) the diurnal variation in top of atmosphere irradiance for the summer and winter solstices.

9.3 Shortwave radiation instruments

9.3.1 *Thermopile pyranometer*

Thermopile pyranometers use a blackened multi-element thermopile as a sensing element, housed beneath a glass hemisphere to protect the sensor from stray thermal

[v] Slight variations (± ~2 W m^{-2}) are known to occur in the sun's output over the 11-year solar cycle, but it is still common to refer to this quantity the 'solar constant', S_0.

Figure 9.8 Pyranometers installed at Reading Observatory, showing the protective glass domes and white body shielding.

losses due to air motion (see Figure 9.8). The properties of the glass domes also define the range of wavelengths transmitted to the sensor, chosen so that little transmission occurs for wavelengths much longer than 3 μm. A thermopile pyranometer's principle of operation is based on detecting the temperature difference between the sensing surface of the thermopile and the instrument's main body to which the thermopile is attached. Such an instrument has a nearly linear response to changes in the incident solar radiation, provided that the inner glass dome has a temperature close to that of the sensor surface.

The thermopile pyranometer is most commonly usually used facing upwards, to measure the global solar irradiance S_g, as the sensor provides a horizontal surface. It can also be used facing downwards to measure the upward (reflected) solar irradiance, S_u, (this configuration is shown in Figure 9.18), but the calibration of instrument may also depend on its orientation (up or down). Finally, if the direct solar beam can be blocked from reaching the pyranometer, for example using a shade ring or a mechanically moved disk, the diffuse solar radiation S_d can also be determined (see Section 9.5).

Figure 9.9 shows global solar irradiance measurements obtained from a horizontally mounted pyranometer, in clear sky and broken cloud conditions, compared with the top of atmosphere radiation calculated for the same day of the year using Equation 9.7. Even in clear sky, the measured irradiance is considerably less than that found from the top of atmosphere calculations because of absorption from atmospheric gases and water vapour. In broken cloud conditions, considerable variability occurs in the surface solar radiation. Some values of solar radiation in these conditions occasionally become greater than the envelope of values might suggest: these transiently large values are associated with forward scattering of the solar radiation at the cloud edge. This 'silver lining' effect is local to the cloud and sensor position.

9.3.2 *Pyranometer theory*

The blackened sensing surface of a thermopile pyranometer can be seen in the foreground pyranometer of Figure 9.8, surrounded by a flat, circular white plate,

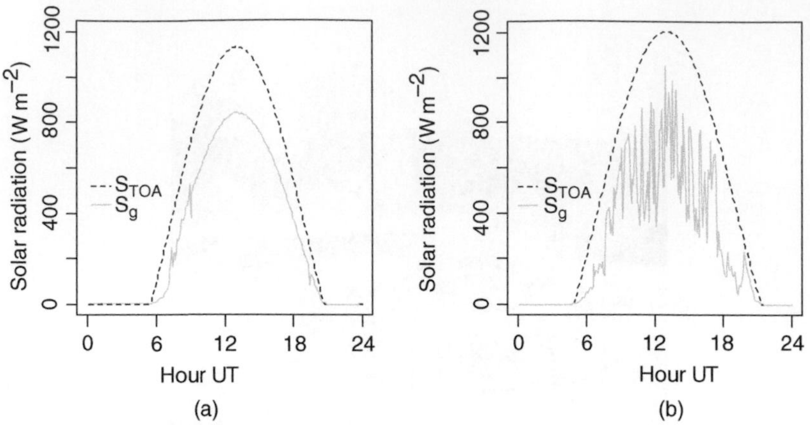

Figure 9.9 Global solar irradiance (S_g,) measured at Reading Observatory compared with the calculated top of atmosphere irradiance (S_{TOA}) for (a) clear sky conditions on 4 August 2007 and (b) broken cloud conditions on 21 June 2006.

shielding the body of the instrument from direct sunshine. The operating principle of the pyranometer can be understood using energy balance arguments.[vi] Consider two sensing elements receiving shortwave radiation through a protective glass dome, one of which is black and the other white (Figure 9.10). The glass dome at temperature T_g also will emit longwave radiation, at a rate $\sigma T_g{}^4$ per unit area, where σ is the Stefan–Boltzmann constant.

For the black surface, the incoming and outgoing energy are in balance when

$$S + \sigma T_g^4 = X_b + \sigma T_b^4, \tag{9.9}$$

and, for the white surface, when

$$\sigma T_g^4 = X_w + \sigma T_w^4, \tag{9.10}$$

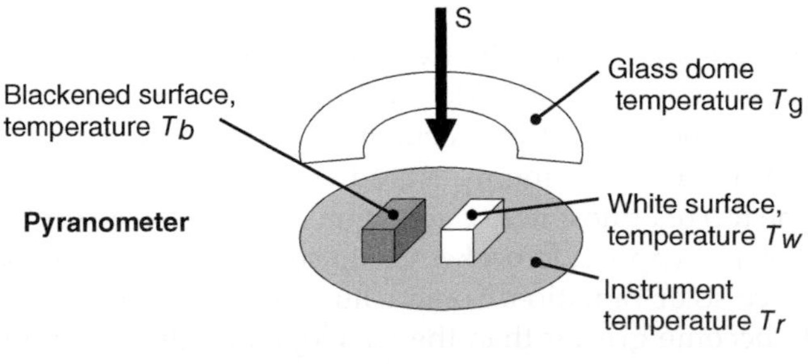

Figure 9.10 Schematic of the energy balance for a simplified pyranometer consisting of two sensing blocks, each of thickness t, one black (which absorbs shortwave and longwave radiation) and one white (which solely absorbs longwave radiation).

[vi] This section follows the approach in W.D. Sellers' Physical Climatology (University of Chicago Press, 1965).

where X_b and X_w are the heat storage rates for the black and white surfaces respectively. Subtracting gives

$$S = (X_b - X_w) + \sigma\left(T_b^4 - T_w^4\right). \quad (9.11)$$

The heat storage rate in each block is, firstly for the black block,

$$X_b = \frac{\kappa}{t}(T_b - T_r), \quad (9.12)$$

and, for the white block,

$$X_w = \frac{\kappa}{t}(T_w - T_r), \quad (9.13)$$

where κ is the thermal conductivity of the material. Taking the difference in the heat storage rates eliminates the reference temperature T_r, as

$$X_b - X_w = \frac{\kappa}{t}(T_b - T_w). \quad (9.14)$$

Approximating $(T_b^4 - T_w^4)$ by $4T_w^3(T_b - T_w)$ and substituting for $(X_b - X_w)$ in Equation 9.11 gives

$$\begin{aligned} S &= \frac{\kappa}{t}(T_b - T_w) + 4\sigma T_w^3(T_b - T_w) \\ &= \left[4\sigma T_w^3 + \frac{\kappa}{t}\right](T_b - T_w). \end{aligned} \quad (9.15)$$

For $4\sigma T_w^3$ small in comparison with $\frac{\kappa}{t}$, this relationship is effectively in the form

$$S = C(T_b - T_w), \quad (9.16)$$

where the thermal properties and sensor thickness can be considered together in a single instrument constant C found through a calibration experiment.

In a practical arrangement, the temperature difference $(T_b - T_w)$ is determined using the thermopile, with T_b the temperature of the blackened sensing surface, and T_w the 'cold' junction temperature. C retains some sensitivity to temperature, and a thermally-uncompensated pyranometer will have a temperature coefficient, that is the emf it produces in response to a fixed irradiance will vary with temperature. This is relatively small (~3% over 5 to 35°C), but, for accuracy, a correction may be needed. In some practical pyranometer instruments, compensation for the temperature coefficient is included directly, using a negative temperature coefficient thermistor (see Section 5.3.3). Such compensation, although in principle non-linear, can reduce the temperature error by at least an order of magnitude [80]. A disadvantage is that the temperature compensation may also increase the output resistance of the instrument, necessitating high input impedance from voltage measuring equipment, if the radiation calibration is not to be affected by circuit loading.

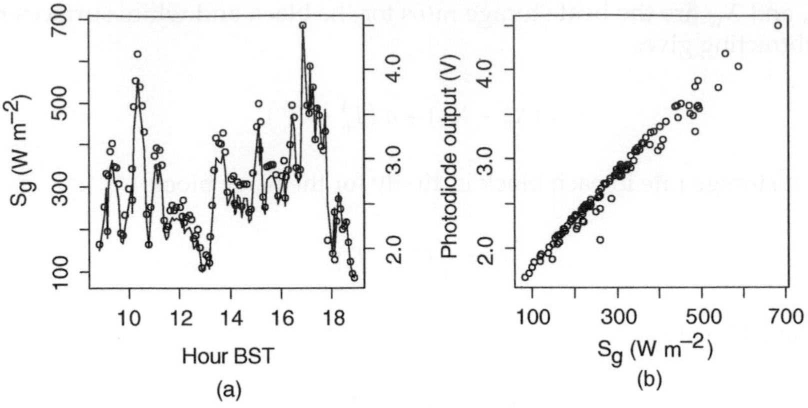

Figure 9.11 Response of a horizontally mounted VTB8440B photodiode to solar radiation, compared with an adjacent S_g measurement obtained using a thermopile pyranometer. (a) Time series of variations in S_g (line) and photodiode output voltage (points) and (b) photodiode voltage plotted against S_g. (Spectral range of photodiode 330 to 720 nm, measurements 5-minute averages from 1-second samples.)

9.3.3 *Silicon pyranometers*

As mentioned in Section 9.1, semiconductors provide another method of radiation sensing. This is based on the sensitivity of a semiconductor to the arrival of photons, which cause a photocurrent to flow. The measurement of photocurrent therefore provides a measurement of the incident radiation, for a range of wavelengths specific to the device considered. Such devices are sometimes referred to as silicon pyranometers, for which the electronic component itself can be very inexpensive.

A disadvantage of a silicon pyranometer is that it has a different spectral response to the well-established broadband response of the thermopile pyranometer, and has a different thermal stability. In these senses at least the silicon pyranometers cannot be directly compared with thermopile pyranometers for high accuracy. However, if only a limited spectral range is required, such as for photosynthetically active radiation (PAR) or ultra-violet measurements, the silicon pyranometer may be preferable. Semiconductor sensors also have a very rapid response, far exceeding that of thermopile devices, which can be an advantage. Although the photocurrent generated is small, it can readily be amplified in a current-to-voltage opamp circuit (Section 3.4.2). Figure 9.11 shows measurements on an inexpensive photodiode used as described in Section 3.4.2, with its sensing surface horizontal. It was exposed to the same solar radiation environment as a thermopile pyranometer device, with both measurements recorded by the same data logging system. It is clear that there is close agreement between the two measurements, although differences emerge at the larger values of solar radiation.

9.4 Pyrheliometers

A pyrheliometer also uses a thermopile sensor, but, because of its need to view the sun at normal incidence, a solar tracking mount is required. Solar trackers are

Figure 9.12 Solar radiation sensors in use at Lerwick Observatory, Shetland. In the foreground and background, four pyranometers can be seen. The two pyranometers in the background are mounted on a solar tracker, shaded by black spheres to allow diffuse radiation measurements. Beneath the two diffuse radiation pyranometers on the solar tracker is a cylinder pointing towards the sun, which houses the pyrheliometer. (The instrument in the foreground on the left-hand side is a sunshine recorder.)

sophisticated mechanical devices, as they are required to move the sensor in both elevation and azimuth. Solar trackers operate by calculating the position of the sun, which may be further refined by direct sensing of the sun's position using a photodiode array.[vii] As the solar tracker requires appreciable area to operate because of the region of space it sweeps out during the year, establishing a small enclosure of radiation instruments is often appropriate (Figure 9.12).

The pyrheliometer measures the direct beam at normal incidence (S_b), leading to a variation with the path length through the atmosphere during which attenuation can occur. Figure 9.13a shows S_b measured at Reading Observatory on an ultra-clear day, without any cloud present. The irradiance at the top of the atmosphere is shown as calculated by Equation 9.7, together with the surface measurements of irradiance made at the surface. The global solar irradiance is the sum of the horizontal component of the direct beam and the diffuse solar radiation, that is

$$S_g = S_b \cos Z + S_d, \tag{9.17}$$

where S_d is the diffuse solar radiation and Z is the solar zenith angle. (Cos Z is calculated for the same time and location as the S_b, S_g and S_d measurements using Equation 9.4.)

Figure 9.13b shows the horizontal irradiance as calculated from the S_b measurement, compared with the measured total horizontal radiation, S_g. On this particularly clear day, S_g and S_b cos Z show close agreement, indicating that the contribution from

[vii] A four-quadrant photodiode uses a circular sensing area having a photodiode in each quadrant. Only if the solar beam falls exactly in the centre of the device, will the output from each of the four photodiodes be almost equal: if one photodiode signal is greater than the others, this information can be used to adjust the solar tracker accordingly.

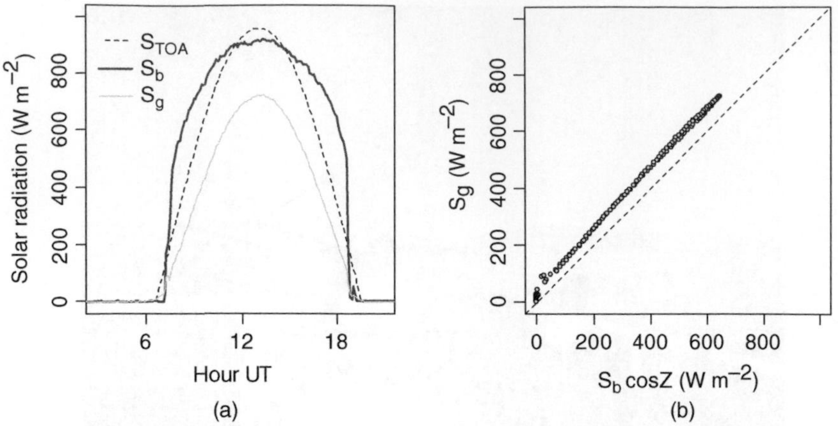

Figure 9.13 (a) Direct beam radiation S_b and global solar irradiance S_g measured at Reading Observatory on an ultra-clear day (7 September 2012). The calculated top of atmosphere irradiance (S_{TOA}) is also shown. (b) S_g plotted against $S_b \cos Z$, where Z is the solar zenith angle calculated for the same time and location.

the diffuse solar radiation (S_d) was small. The diffuse radiation S_d can be derived by differencing such measurements, but, rather than require two well-calibrated instruments and a solar tracker, single instrument methods for determining S_d directly can instead be employed.

9.5 Diffuse solar radiation measurement

In principle, measurement of the diffuse radiation requires nothing more than a sensor which is shaded from the direct solar beam. If the sky is completely overcast, no shading is required as the radiation received at the ground will be almost completely diffuse, with, in the ideal case, the radiance uniform across the whole sky.[viii] Under a cloudless sky, however, the diffuse radiation is not straightforwardly characterised. Compared with the average across the entire sky, the region of the sky around the sun is several times brighter, falling to about half the whole sky average by about 90° from the sun also depending on atmospheric aerosol loading.

Practical methods for obtaining a measurement of the diffuse radiation in cloudless or partially cloudy conditions are now considered further.

9.5.1 *Occulting disk method*

One method of preventing the direct radiation from reaching a sensor is to position an object (known as an *occulting disk*) to shade it. The position of an occulting disk will need to be constantly adjusted because of relative motion in the sun-earth system, if the direct beam is to be continuously obscured. For brief calibrations, this can

[viii]This is known as a *Uniform Overcast Sky (UOC)*. The radiance of a real overcast sky can be calculated at any zenith angle using assumptions for a *Standard Overcast Sky (SOC)*.

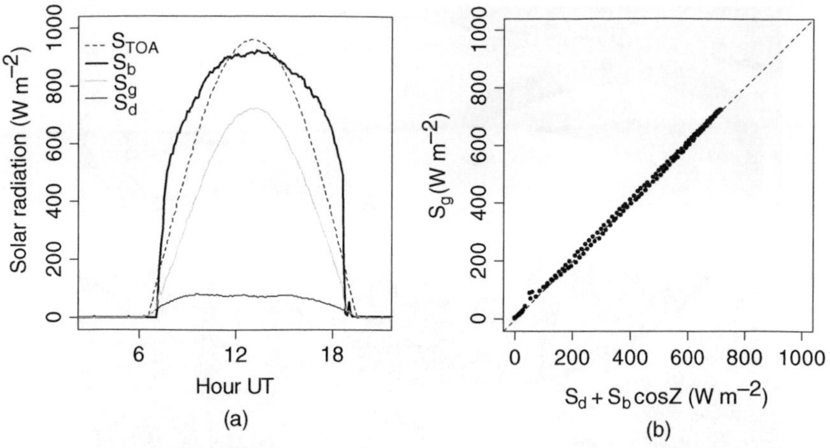

Figure 9.14 (a) Solar radiation measurements made at Reading on 7 September 2012 (as for Figure 9.13) but with the measurement of the diffuse solar radiation (S_d) added. (b) S_g plotted against ($S_d + S_b \cos Z$), where Z is the solar zenith angle calculated for the same time and location.

be done by hand, but this is unsatisfactory for continuous measurements and a solar tracker is required. By mounting the occulting disk (or sphere) on the solar tracker, the direct beam can be constantly obscured from reaching a pyranometer. (An example of the practical use of occulting spheres for the diffuse measurement is given in Figure 9.12.) Using the occultation measurement technique, the variation in diffuse radiation determined on 7 September 2012 has been included in Figure 9.14. The close agreement in 9.14b from three independently obtained measurements demonstrates the validity of Equation 9.17.

9.5.2 *Shade ring method*

A simpler method for obtaining the diffuse radiation is to place the pyranometer within a circular band (or *shade ring*) designed, positioned and regularly adjusted to prevent the direct solar beam reaching the sensor. This technique is in widespread use, but has the disadvantage that, as well as obscuring the direct solar radiation, part of the diffuse radiation sought is also lost due to interception by the shade ring. The amount of diffuse radiation intercepted varies with latitude and time of year, and a correction factor is needed. Figure 9.15 shows a shade ring fitted to a pyranometer.

The diffuse radiation lost because of the presence of the shade ring can be corrected by multiplying by a factor k, given by

$$k = \frac{1}{1 - qf}, \tag{9.18}$$

where f arises from the solar-terrestrial geometry and q is an anisotropy factor varying between 0 and 1 to allow for the effect of varying brightness in the sky around the sun under non-overcast conditions. (The simplest case of an isotropic assumption fails

Figure 9.15 The use of an adjustable shade ring with a pyranometer to prevent the direct beam reaching the sensor.

under thin or broken cloud [81] when there is a bright region surrounding the sun which contributes circumsolar radiation.[ix])

The geometrical factor f describing the obscuration by the shade ring varies with time of year and is given by [82]

$$f = \left(\frac{2}{\pi}\right)\left(\frac{b}{r}\right)(\cos^3\delta)\,[t_{set}\sin d\sin\phi + \sin(t_{set})\cos\delta\cos\phi]\,. \qquad (9.19)$$

Here, (b/r) is the width-to-radius ratio of the shade ring, t_{set} is the hour angle for sunset and δ the solar declination for the day of the year concerned and ϕ is the latitude. The variation of the correction factor k is calculated as a function of δ in Figure 9.16, assuming isotropic sky conditions when completely overcast ($q = 1$). There is some seasonal asymmetry apparent, in that, in the summer (largest solar declination), the sun is approximately overhead for a larger proportion of the day compared with the winter, and the position of the shade ring required removes a greater proportion of the diffuse radiation than in the winter.

Co-located measurements of the diffuse radiation using both the occulting disk and shade ring methods allow the effectiveness of the geometrical correction calculation to be evaluated. Figure 9.17 shows simultaneous diffuse radiation measurements obtained using the shade ring and occultation methods. Assuming that the occultation method provides a 'true' diffuse radiation measurement, the ratio between the occultation and shade band diffuse radiation can be expected to be dominated by the effect of the shade band. Whilst some days show considerable variability which may be site-related, the seasonal variation of the ratio between the method does largely follow that expected from Equation 9.19.

[ix] Additionally, q may also show a variation with the solar angle, because of local atmospheric turbidity.

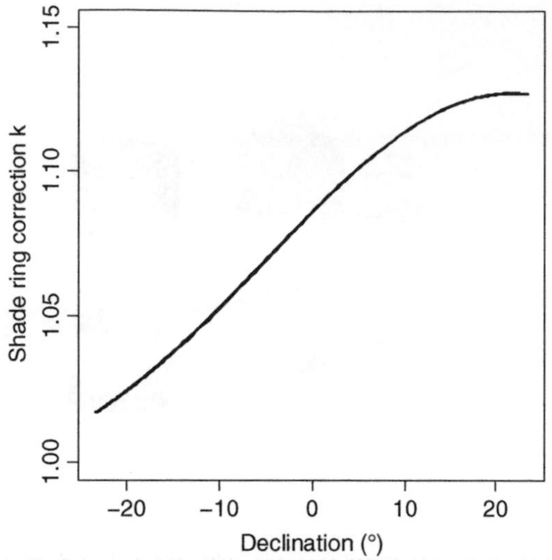

Figure 9.16 Shade ring correction k, plotted as a function of the solar declination, for a shade ring width–radius ratio of $b/r = 0.2$ at the latitude of Reading (51.442°N).

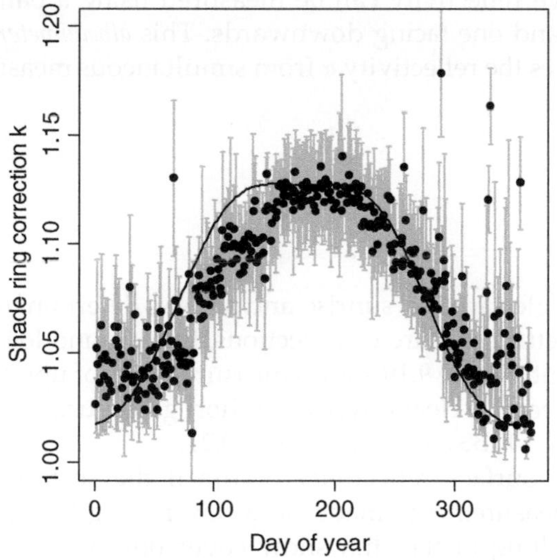

Figure 9.17 Shade ring correction k for the diffuse radiation calculated from Equation 9.18, using the isotropic sky assumption ($q = 1$) and a shade width–radius ratio of $b/r = 0.2$, for the latitude of Reading (51.442°N). Points show the ratio of the measured occulted disk diffuse radiation to the shade band radiation at Reading during overcast periods (identified by $S_d/S_g > 0.8$), for daily mean values between January 2011 and December 2012. (Error bars show the daily standard deviation of the ratio.)

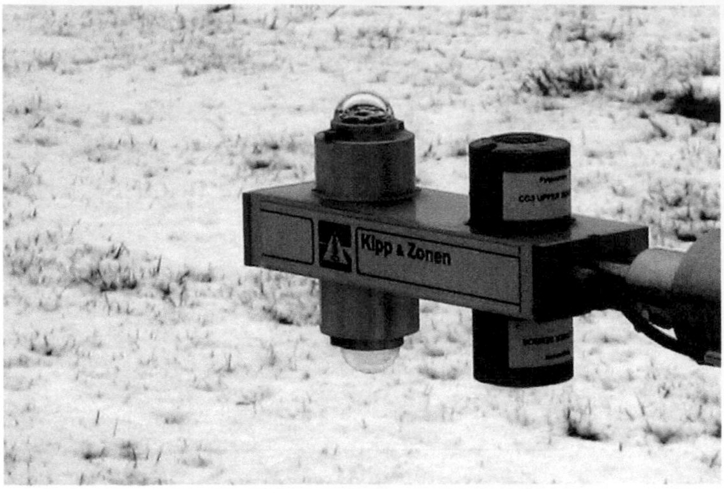

Figure 9.18 The combination of two pyranometers, with their glass domes oriented upwards and downwards, allows the shortwave reflectivity of the surface beneath to be determined.

9.5.3 *Reflected shortwave radiation*

The surface shortwave reflectivity can be measured using a pair of pyranometers, one facing upwards and one facing downwards. This *albedometer* arrangement (see Figure 9.18) determines the reflectivity α from simultaneous measurements of S_g and S_u, as

$$\alpha = \frac{S_u}{S_g}. \tag{9.20}$$

For the low sun angles around sunrise and sunset, determinations of α are unreliable because of the effect of glare or reflections, but, around local noon, the values become very consistent. Figure 9.19 shows measurements of upward and downward solar radiation obtained on a clear day, over a short grass surface. The average reflectivity obtained between 12BST and 14BST was 0.21.

A large change in surface reflectivity occurs if the surface has snow cover.[x] Figure 9.20 shows measurements made on a day having had light snowfall from 04 to 05UT, with melting of the thin snow cover observed from about 11UT. In Figure 9.20a, the S_d and S_g measurements remain similar throughout the day, indicating overcast conditions, but the reflected shortwave radiation falls sharply around the time the snow was observed to melt. This is unlikely to be due to the effect of low sun angle, as, during the same time interval for the days before which had no snow cover, there was no such enhancement.

[x] In determining reflectivity over snow, it is essential to ensure that any snow cover on the upward facing pyranometer has been cleared, as this will affect the calculated reflectivity through the numerator of Equation 9.20.

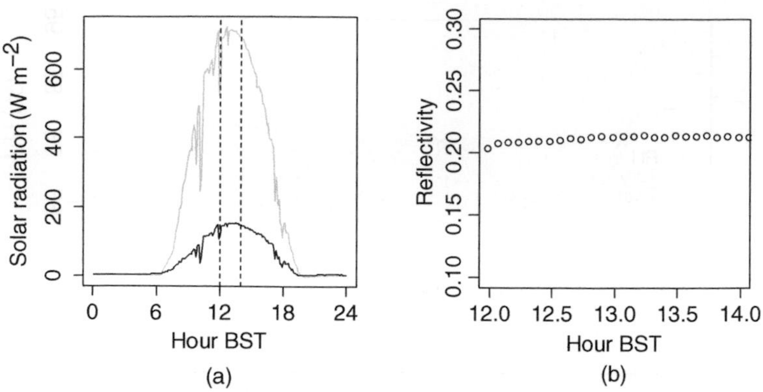

Figure 9.19 (a) Solar radiation measured on 4 September 2012 over the short grass of the Reading Observatory site, showing S_g (grey line) and S_u (black line) as 5-minute averages from 1-second samples. The hours around local noon are marked with dashed vertical lines. (b) Shortwave reflectivity (S_u/S_g) calculated for the hours around local noon.

9.5.4 *Fluctuations in measured radiation*

The direct beam is readily attenuated by the presence of cloud, and S_b is greatly reduced under cloud of any appreciable optical thickness. The solar radiation at the surface then becomes principally diffuse. Figure 9.21 shows solar radiation measurements during the transition from negligible thin cloud to substantial thick cloud. Soon after 12UT on this day, S_g and S_d became very similar, indicating overcast conditions. Correspondingly, S_b dropped to close to zero at the same time. However, the transition to overcast on this day began with the passage of wave cloud, photographed at 1130UT in Figure 9.22, which demonstrates the evolution in attenuation of the direct beam.

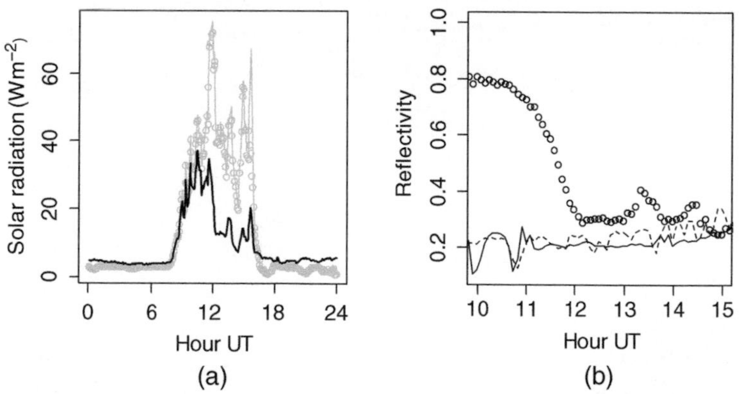

Figure 9.20 Effect of melting show on reflected shortwave radiation. (a) Solar radiation measured at Reading Observatory on 14 January 2013, showing S_g (grey line), S_d (grey points) and S_u (black line). (b) Shortwave reflectivity (S_u/S_g) for the data from 14 January 2013 (points), and also calculated from the same instruments for 13 January 2013 (solid line) and 15 January (dashed line). (All instruments were inspected at 09UT.)

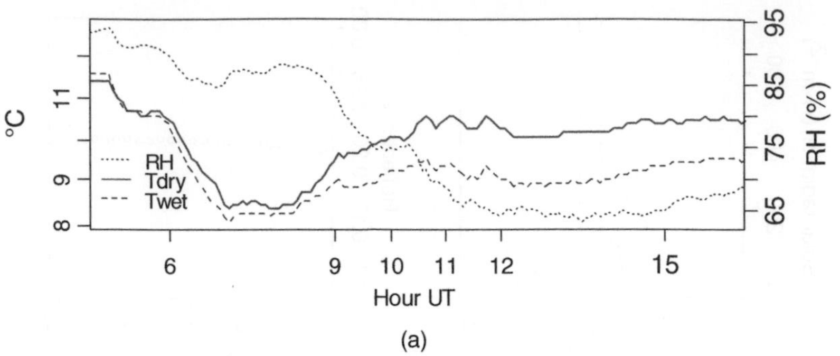

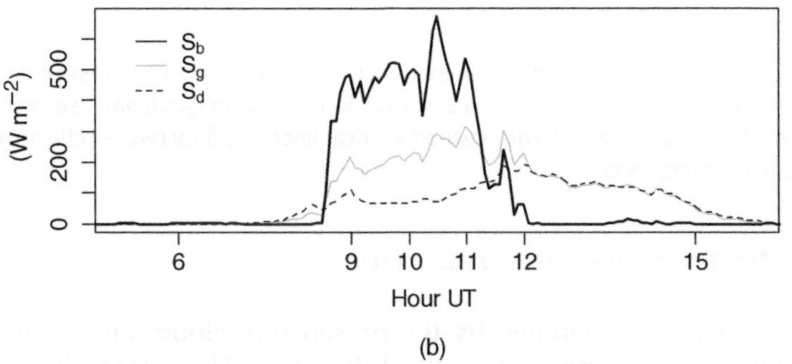

Figure 9.21 The effect of cloud on the direct solar beam on 11 November 2006 at Reading. (a) Surface temperature and humidity measurements. (b) Direct solar beam (S_b), diffuse solar radiation (S_d) and global solar irradiance (S_g).

Figure 9.22 Wave cloud observed from Reading on 11 November 2006 at 1130UT.

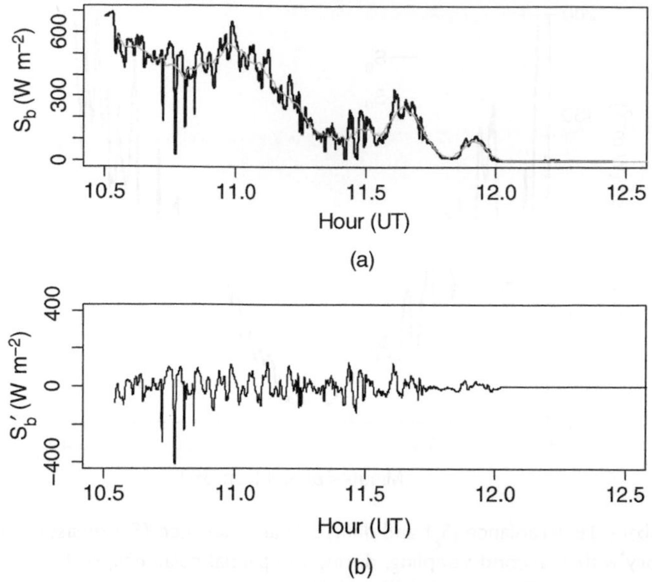

Figure 9.23 High temporal resolution direct beam solar radiation measurements from 11 November 2006 (Figure 9.21), showing (a) the raw data (10-second averages) with a slower (5 min) moving average calculated (grey line). (b) Direct beam fluctuations (S_b') determined by subtracting the moving average in (a).

Wave cloud propagation can occasionally be detected in the direct beam when the cloud is sufficiently optically thin for the direct beam to be partially attenuated, with rapid periodic variations. Figure 9.23a shows small rapid fluctuations evident in the direct beam beneath the wave clouds on 11 November 2006, which occur on a shorter timescale in comparison with other more substantial changes occurring. By smoothing out the longer timescales which are then subtracted, improved resolution of the rapid fluctuations can be achieved (see Figure 9.23b). There is a change in character of the fluctuations from about 1130UT as S_b begins to become substantially attenuated with the cloud's increasing optical thickness until the direct beam is entirely cut off.

Further high-temporal resolution measurements of solar radiation are shown in Figure 9.24. The pyranometer response time removes variations occurring more rapidly than about 10 s, but even so, on the tens of seconds timescales the diffuse radiation shows much less variability than the global solar radiation (see also Figure 12.20).

9.6 Reference solar radiation instruments

Calibration of solar radiation instruments ultimately requires an absolute instrument, which can be operated under controlled conditions, but stable secondary instruments, able to transfer the calibration from the absolute instrument to instruments in regular use are also required. For radiometry, the cavity radiometer is the absolute instrument.

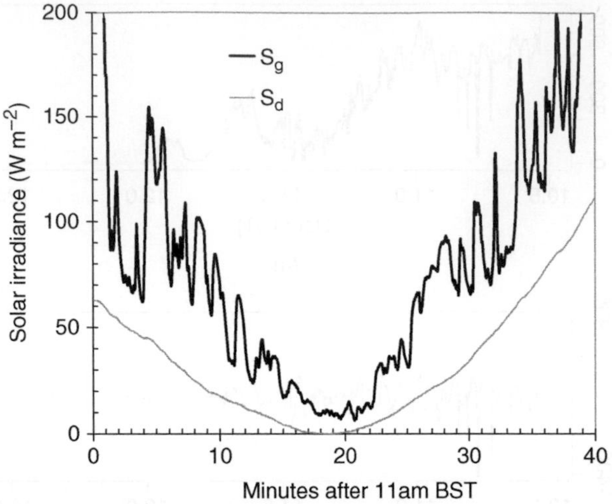

Figure 9.24 Global solar irradiance (S_g) and diffuse solar irradiance (S_d) measurements obtained at Reading Observatory with 1-second sampling, during the partial solar eclipse [83] (11 August 1999), with the maximum eclipse of 97%, at 1118 BST and 41 s.

9.6.1 *Cavity radiometer*

The cavity radiometer is a primary standard radiometer. It uses a blackened cavity containing a blackened metal cone to absorb the solar radiation falling on it. The base of the cone is connected to the instrument's body through a thermal resistance. Consequently, heat absorbed by the cone from solar radiation flows produces a temperature difference proportional to the heat flow. By blocking off the solar radiation, the equivalent quantity of electrical heating necessary to give the same heat flow through the thermal resistance (generating the same temperature difference) can be found. Following this comparison, if the area A of the surface receiving solar irradiance S_b at normal incidence is known precisely, S_b can be calculated.

The cavity radiometer is an absolute instrument, used for calibration of all other shortwave radiometers, pyrheliometers and pyranometers.

9.6.2 *Secondary pyrheliometers*

Secondary pyrheliometers are calibrated from the cavity radiometers, for use as transferable calibration instruments via their calibration certificate obtained by comparison with a cavity radiometer. The Linke–Fuessner design of pyrheliometer provides such a secondary standard device. This uses a Moll–Gorczynski thermopile as the sensor, housed in a heavy metal tube. It has manual controls for elevation and azimuth, to allow it to be used to follow the sun to measure the direct beam solar radiation at normal incidence. The field of view of the instrument is about 5°, with the incoming radiation collimated by a series of baffles in the tube. Optional filters allow selection of different wavebands.

9.7 Longwave instruments

The net radiation at the surface has longwave and shortwave contributions, as given in Equation 9.1. Methods to measure the broadband longwave radiation are now considered, which, in terms of sensing, show considerable similarities with the shortwave measurements. The detailed operation of a pyrradiometer provides an example.

9.7.1 *Pyrradiometer theory*

As for a pyranometer, a pyrradiometer's operation can be understood in terms of the energy balance established on its upper and lower sensing surfaces, shown schematically in Figure 9.25. The exposed upper and lower sensing elements are mounted horizontally and separated by a thin insulating layer of material of thickness *d*.

Consider the energy balance for the upper element in thermal equilibrium. Down-welling shortwave (S_g) and longwave (L_d) radiation is balanced by the heat storage rate (X_u), convective heat loss rate H_u and radiative loss rate σT_u^4,

$$S_g + L_d = X_u + H_u + \sigma T_u^4. \tag{9.21}$$

For the lower element, the up-welling shortwave can be found from the reflected shortwave (αS_g) as

$$\alpha S_g + L_u = X_d + H_d + \sigma T_d^4. \tag{9.22}$$

Combining these two equations gives

$$S_g(1-\alpha) - L_u + L_d = X_u - X_d + H_u - H_d + \sigma\left(T_u^4 - T_d^4\right), \tag{9.23}$$

which can be recognised from Equation 9.1 as the net radiation, R_n.

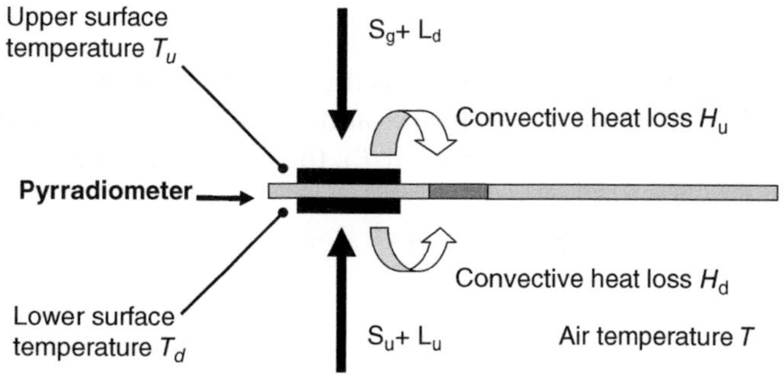

Figure 9.25 Schematic of the energy balance for the two sensing surfaces of a pyrradiometer. (Some heat is also stored in the sensing surfaces and body of the instrument.)

The net storage rate ($X_u - X_d$) is related to the temperature difference between the upper and lower sensors as

$$X_u - X_d = \frac{\kappa}{d}(T_u - T_d), \quad (9.24)$$

for κ the thermal conductivity of the insulating material. The convective heat loss rates H_u and H_d are related to the temperature difference between each sensing surface and its surroundings by

$$H_u \propto (T_u - T), \quad (9.25)$$

and

$$H_d \propto (T_d - T), \quad (9.26)$$

giving, for convective losses varying with wind speed u as $f(u)$,

$$H_u - H_d = f(u)(T_u - T_d). \quad (9.27)$$

Substituting for these terms in Equation 9.23 and approximating $(T_u^4 - T_d^4)$ by $4T_d^3(T_u - T_d)$ gives

$$R_n = \left[\frac{\kappa}{d} + f(u) + 4\sigma T_d^3\right](T_u - T_d). \quad (9.28)$$

This shows that the net radiation R_n is related to $(T_u - T_d)$ if the terms in the square brackets can be assumed to be constant, which amounts to assuming that the ventilation rate u is constant and the $4\sigma {T_d}^3$ term varies little. As for a pyranometer, the temperature difference $(T_u - T_d)$ is readily measured using a thermocouple, thereby providing a voltage output proportional to R_n.

9.7.2 *Pyrradiometer calibration*

A method for calibration of a pyrradiometer is illustrated schematically in Figure 9.26, using two thermally stable radiating surfaces above and below. For the upper surface at a temperature T_u and the lower surface at T_d, the net radiation R_n is

$$R_n = \sigma\left(\varepsilon_u T_u^4 - \varepsilon_d T_d^4\right), \quad (9.29)$$

where σ is the Stefan–Boltzmann constant (5.67×10^{-8} W m^{-2} K^{-4}) and ε_u and ε_d are the upper and lower surface emissivity respectively. Assuming unit emissivity for both surfaces ($\varepsilon_u = \varepsilon_d = 1$), R_n received by the instrument can be calculated, as T_u is varied.

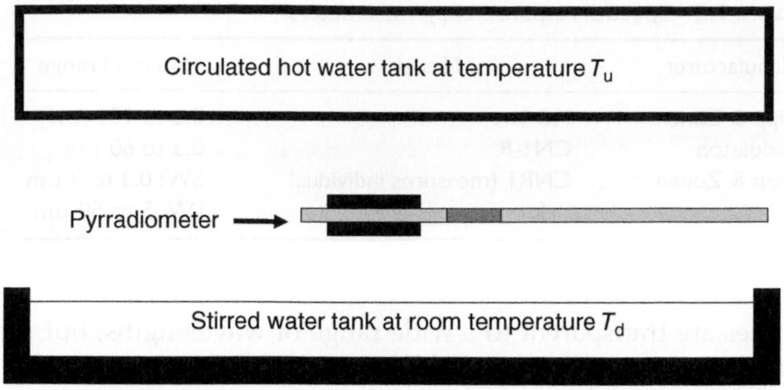

Figure 9.26 Principle of calibration for a pyrradiometer, using radiation emitted from extensive water tanks above and below the sensor. The temperature of the upper tank is varied to obtain a range of radiation values with the lower tank temperature held constant, whilst the output voltage from the pyrradiometer is recorded.

9.7.3 *Pyrgeometer measurements*

A pyrradiometer can be used solely to determine longwave radiation if the measurements are made nocturnally or if the shortwave term is independently measured and subtracted. This can be illustrated by considering containing the lower sensing surface within an internally blackened aluminium cup, which radiates at a known temperature, either assumed to be the air temperature T or directly measured. This is shown in Figure 9.27. Equation 9.28 can then be modified slightly to give

$$S_g + L_d = \left[\frac{\kappa}{d} + f(u) + 4\sigma T_d^3\right](T_u - T_d) + \sigma T^4, \qquad (9.30)$$

showing that, if S_g is also measured (or zero nocturnally), L_d can be found.

9.7.4 *Commercial pyrradiometers*

Many commercial versions of pyrradiometers exist, and the responses of some are summarised in Table 9.2. The Middleton design of net radiometer uses a blackened, multi-element thermopile with two parallel faces, protected by hemispheric domes of thin polythene to reduce convective losses from the sensing surfaces (see Figure 9.28).

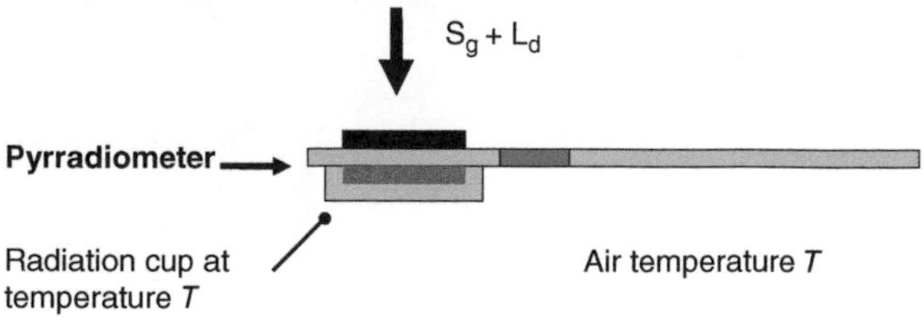

Figure 9.27 Energy balance for a pyrradiometer with a radiation cup fitted to the lower sensing surface.

Table 9.2 Spectral response of pyrradiometers

Manufacturer	Model	Spectral range
Kipp & Zonen	NR Lite	0.2 to 100 μm
Middleton	CN1-R	0.3 to 60 μm
Kipp & Zonen	CNR1 (measures individual four components)	SW: 0.3 to 3 μm LW: 5 to 50 μm

Polythene domes are transparent to a wide range of wavelengths, but they require inflation by dry air or nitrogen and therefore a pump or gas supply which may be difficult to arrange.

In the Kipp and Zonen NR Lite net radiometer, shallow cones of Teflon are used instead of the more conventional flat sensor plates, which avoids the need for inflation of polythene domes to cover the sensing surfaces. A disadvantage of this arrangement is that the response is affected by natural ventilation.

Net radiometers are nearly always mounted about 1 m above the ground with the sensor surfaces parallel to the ground. The instrument then monitors a net flux density which represents the effective atmospheric radiation absorbed by the Earth's surface. The energy balance arguments presented (Section 9.7.1) indicate such an instrument should have a nearly linear response to changes in the net radiation, if the temperature difference between the two sensor surfaces is not too large. Dew (or frost) on the outside of the domes invalidates the readings: preventative measures include the use of heating wires or jets of dry air.

The net radiation R_n shows, in daytime, a similar form to the solar radiation as the solar contribution is generally the largest term in Equation 9.1. Figure 9.29 shows the variation in global solar irradiation (S_g), measured using a pyranometer on a clear day, and the same day's variation in net radiation (R_n), from an adjacent

Figure 9.28 Use of pyraddiometers (and a downward facing pyranometer) at a remote site in the Sahel. The pyrradiometers use upward and downward polythene domes for protection, inflated by dry air from a pump.

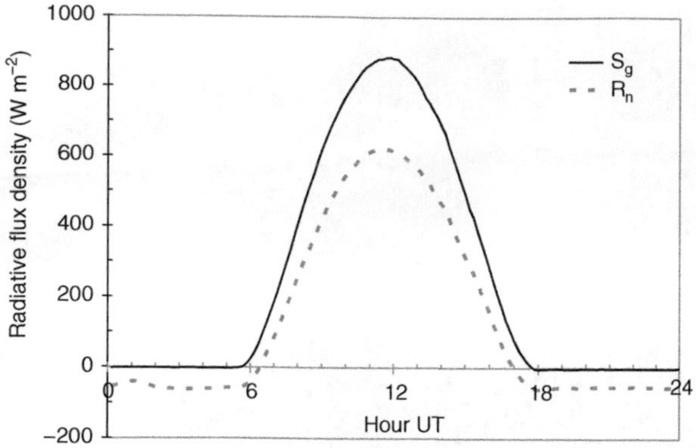

Figure 9.29 Global solar radiation (S_g) and net radiation (R_n) measured during a cloudless day (25 September 1992) in the Sahel (the soil temperature measurements made nearby on the same day are shown in Figure 5.25).

pyrradiometer. During the day, R_n is positive; at night, R_n becomes negative, as heat is released from the soil.

A combination of separate shortwave and longwave sensors is used in the Kipp and Zonen CNR1 pyrradiometer (Figure 9.30), which allows the individual radiation components to be measured and the net radiation to be derived. A correction is needed for radiation emitted at the instrument body temperature, which is measured using an internal resistance thermometer. The longwave measurements provide further information on the cloud beyond that available from shortwave measurements, notably at night, but also, approximately on the cloud height. Low, warmer cloud emits more downwelling longwave radiation than high-level ice cloud or a clear sky. There can consequently be sharp transitions in the downwelling longwave associated with changes in cloud.

Figure 9.31 shows time series of the four components of radiation measured on a sequence of clear, overcast and broken cloud days. It can be seen that during the day, the solar radiation dominates. Towards the end of day 44, the transition from clear to overcast conditions is apparent as L_d increases (which is also apparent in R_n), and then remains steady through the overcast day. The variation in incoming shortwave (S_g) is considerably suppressed by the presence of the cloud on day 45. During day 46, the cloud breaks, and L_d decreases sharply. S_g also becomes more variable as the cloud breaks. In Figure 9.31b the two methods of determining the net radiation are compared, firstly using a pyrradiometer, and secondly by calculation from measuring the four components separately. The two approaches are consistent. In Figure 9.31c, the air temperature variation lags S_g on day 44, and remains fairly steady under the overcast conditions until the cloud breaks on day 46.

9.7.5 *Radiation thermometry*

The variation in emitted longwave radiation with the temperature of the emitting object provides a method of remote temperature measurement. A radiation

Figure 9.30 A four-component net radiometer, measuring the upwelling and downwelling shortwave radiation (domed sensors), and the upwelling and downwelling longwave radiation.

thermometer ('IRT'[xi]) measures infrared radiation emitted in a narrower part (or 'window') of the electromagnetic spectrum compared with a broadband longwave measurement for example between 7.5 to 13.5 μm. From the radiation received, the temperature of the emitting surface can be calculated, using the Stefan–Boltzmann relationship, with a correction for the surface's emissivity. When pointed at the sky, an IRT does not measure local air temperature, but a weighted average temperature from the atmospheric sources of radiation which are emitting in its wavelength window. The typical resolution for an IRT is 0.1°C, but, because of uncertainty in the emissivity or atmospheric absorption, the absolute accuracy is poorer, about 2°C depending on the temperature.

9.8 Sunshine duration

Although not a measurement of the amount of radiation, recording the duration of bright sunshine is a measurement made at many meteorological and climatological sites. With the calculated day length from Equation 9.6, the maximum sunshine duration possible at any site for any day can be found, and the daily sunshine

[xi] InfraRed Thermometer. This should not be confused with a *black bulb thermometer*, which, and as its name implies, is designed to absorb longwave radiation. Its practical use is complicated by the need to correct for ventilation effects.

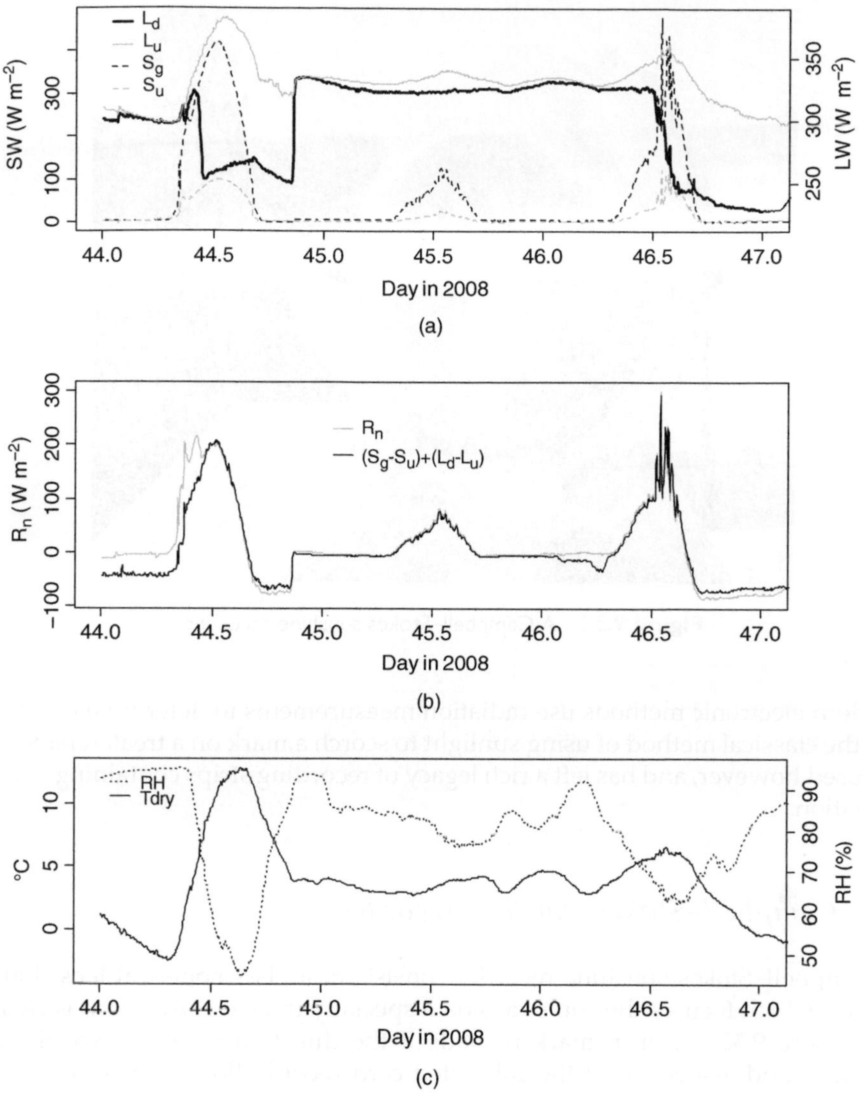

Figure 9.31 (a) Time series of downwelling and upwelling longwave (LW; L_d, L_u, right-hand axis) and shortwave (SW; S_g, S_u, left-hand axis) radiation components for a clear day (day 44), an overcast day (day 45) and a broken cloud day (day 46), as measured at Reading Observatory. (b) Net radiation (R_n), measured using a pyrradiometer (grey line) and derived from the four components measured separately (black lines). (c) Simultaneous air temperature (T_{dry}) and relative humidity (RH). In (a), the incoming radiation is marked with the black lines, outgoing in grey.

duration determined as a fraction of the maximum duration possible. Deriving the daily sunshine fraction provides a normalisation to allow comparison of sunshine durations at different times of year.[xii]

[xii] The sunshine fraction can be used with the *diffuse fraction* and *clearness index* to give bulk measures of cloud scattering and atmospheric transmission. (For their definitions see Section 12.2.2).

Figure 9.32 A Campbell–Stokes sunshine recorder.

Modern electronic methods use radiation measurements to determine bright sunshine; the classical method of using sunlight to scorch a mark on a treated paper strip is still used however, and has left a rich legacy of recording strips containing sky state information.

9.8.1 *Campbell–Stokes sunshine recorder*

The Campbell–Stokes sunshine recorder consists of a glass spherical lens, 100 mm diameter, which focuses the sun's rays on a specially treated card which is changed daily (Figure 9.32). A burn mark is made if the direct solar beam exceeds about 120 W m^{-2} and inspection of the calibrated card record allows a determination of 'hours of bright sunshine' for the day concerned (Figure 9.33). It is useful in climatology, as no electrical power is needed, and daily totals are easily obtained. However,

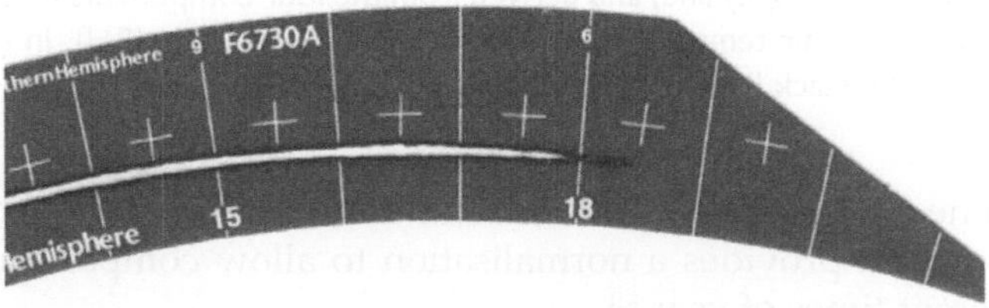

Figure 9.33 Marks on the burn card from the Campbell–Stokes sunshine recorder at Reading, for 11 May 2008.

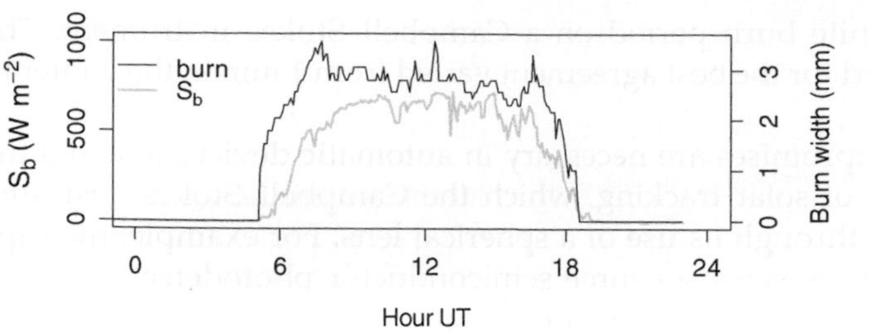

Figure 9.34 Comparison of the direct beam radiation S_b (grey line) measured at Reading Observatory, with the width of the Campbell–Stokes burn trace (black line) obtained at the same location, for 11 May 2008.

it does not provide a measure of total energy received, merely an indication of the time for which the instrument's burn threshold was exceeded.

The Campbell–Stokes sunshine recorder is arguably one of the most elegant meteorological instruments for continuous environmental measurement, but, unfortunately, it is also one of the least useful because of its poor quantitative response. It is inexact primarily because the burn threshold can vary, for example between 106 and 285 W m^{-2} of direct beam irradiance [84], and the burn may persist during intermittent bright sunshine, leading to an overestimation of the sunshine duration. The burn threshold may also depend on the dampness of the card, leading to a difference of about 8% between sunset and sunrise [85]. An additional factor is that the subjective interpretation of intermittent burn marks may vary. However, the summed burn duration from this device has been obtained at many sites for a long time and in some cases over a century.

The actual physical record of the sky provided may now become more important as retrieval methods improve [86]. This is because, as well as the length of the burn mark providing the sunshine duration, it has long been recognised that there is also information contained in the width of the burn, which is proportional to the intensity of the direct irradiance [87]. Figure 9.34 shows a comparison between the width of the burn mark and S_b, for a day of bright sunshine [88].

9.8.2 *Electronic sensors*

Electronic methods have also been developed for measuring sunshine such as the Foster-Foskett sunshine switch developed in the 1950s and used in the USA. Electronic devices measure the absolute level of solar radiation and test whether it exceeds a threshold value. A related approach is to establish when the diffuse radiation received by an unshaded sensor is the same as that at a shaded sensor. When the signals from both sensors are the same, the sky is assumed to be overcast. Conversion of the measurements obtained to mimic the Campbell–Stokes instrument requires an algorithm to emulate the response of the Campbell–Stokes instrument. In a study using an electronic instrument employing a detector array of photodiode sensors [89], agreement to 0.5% in the annual sunshine duration was obtained if each instantaneous bright sunshine measurement above a threshold was assumed to represent a finite burn period on a Campbell–Stokes instrument. The finite burn period required for the best agreement varied from 1 min in the winter to 6 min in the summer.

Several compromises are necessary in automatic devices to avoid the mechanical complications of solar tracking, which the Campbell–Stokes' instrument elegantly circumvented through its use of a spherical lens. For example, the Kipp and Zonen CSD1 sunshine sensor uses three semiconductor photodetectors, one (D1) to measure the all sky diffuse and direct radiation and two further sensors (D2 and D3) to measure different regions of the sky. The instrument determines which of D2 and D3 receives the smaller signal, and assumes that this represents one-third of the diffuse radiation. The value of D1 is then corrected for the measured diffuse radiation, from which the direct radiation can be found. If the direct radiation exceeds 120 W m^{-2}, the conditions are recorded electronically as 'bright sunshine'. The uncertainty quoted by the manufacturer is $\pm$ 40 W m^{-2}. (A CSD1 sunshine sensor is shown in the foreground of Figure 9.12.)

10

Clouds, Precipitation and Atmospheric Electricity

10.1 Introduction

The state of the sky and perception of visibility have influenced artists [90], writers and even musicians [91] for centuries, but important practical implications for air and road transport exist because of the hazards associated with fog and heavy precipitation. Clouds and fogs also strongly perturb the atmospheric electrical environment from its quiescent state in fair weather. The discussion here is restricted to *in situ* measurements and therefore mostly to cloud-related properties which can be sensed from the surface with basic instruments. (Some effects of clouds on surface radiation measurements are considered in Section 12.2.2.)

10.2 Visual range

The visual range V is the greatest distance at which an object can be seen and identified by an observer with normal vision under normal daylight conditions. Reduction in the maximum visual range is caused by scattering of light by water droplets or aerosol particles, and also by very slight absorption. The visual range at a meteorological station is usually assessed against objects (or lamps during nighttime) at fixed distances, and reported in a coded form.

For visible light, the important scattering particles are relatively large with respect to the wavelength (often droplets with micron diameters or greater), hence the scattering is more or less independent of wavelength. For a parallel beam of irradiance I passing through a thickness dx of scattering material, the reduction in irradiance dI is given by Beer's law as

$$dI = -\xi I(x)dx, \tag{10.1}$$

where ξ is an extinction coefficient. If ξ is independent of x, then

$$I = I_0 \exp(-\xi x). \tag{10.2}$$

Meteorological Measurements and Instrumentation, First Edition. R. Giles Harrison.
© 2015 John Wiley & Sons, Ltd. Published 2015 by John Wiley & Sons, Ltd.
Companion website: www.wiley.com/go/harrison/meteorologicalinstruments

Assuming an extinction cross section for droplets which is twice their projected area,[i] the total extinction coefficient, when there are N drops per unit volume of diameter D, is, approximately,

$$\xi = N\frac{\pi}{2}D^2 . \tag{10.3}$$

To determine the visual range in a way which is consistent with human observations, the extinction coefficient has to be related to the physiological response of the human eye. Koschmieder's theory [92] relates the visual contrast between an object and its surroundings to the light-scattering properties of the intervening atmosphere. For a dark object at range X viewed against a light background, the visual contrast C at an observer's eye is

$$C = \exp(-\xi X) . \tag{10.4}$$

As X increases, the contrast C between the dark object and the light background decreases, until C reaches a critical value, ε. This is generally taken [93] to be $\varepsilon = 0.05$. At this critical value, the object can no longer be distinguished from its surroundings and the object is then at the visual range V. Thus

$$\varepsilon = \exp(-\xi V) , \tag{10.5}$$

and the visual range is

$$\begin{aligned} V &= -\frac{\ln \varepsilon}{\xi} \\ &= \frac{6}{\pi N D^2} . \end{aligned} \tag{10.6}$$

Figure 10.1 shows calculations of visual range based on Equation 10.6, for a range of assumed droplet sizes and concentrations.[ii]

10.2.1 *Point visibility meters*

These instruments (also known as forward scatter meters) measure the scattering of light directly, from a narrow collimated beam at a precisely-defined angle to the beam (usually about 50°), within an optical instrument allowing ingress of fog droplets. Again the scattering measurements allow calculation of ξ, from which the visual range V is estimated. A disadvantage is that the small sample volume may give unrepresentative readings under some conditions.

---

[i] See, for example, Bohren, C.F. and Huffman, D.R. 1983. *Absorption and Scattering of Light by Small Particles*. John Wiley & Sons, Inc., New York.

[ii] An alternative formulation for nocturnal conditions is provided by Allards' Law, originally devised to calculate the operating range of lighthouses.

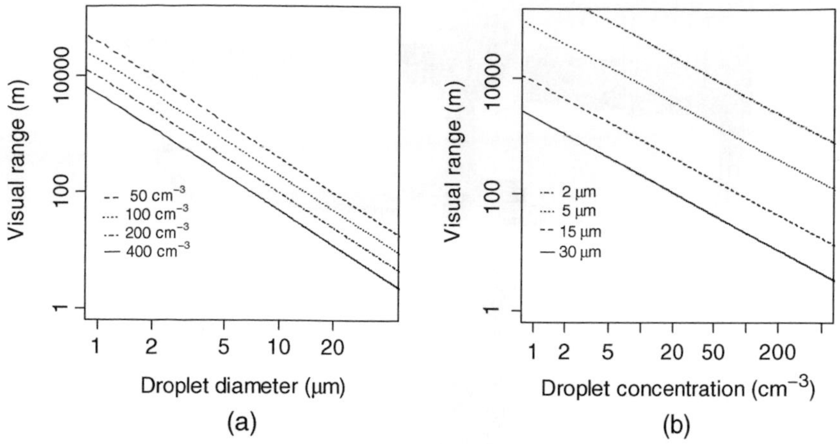

Figure 10.1 Visual range calculated as a function of (a) droplet diameter and (b) droplet concentration. (In each case the droplets are all assumed to have the same radius.)

10.2.2 *Transmissometers*

Rather than the manual determination of an observed visual range, instruments known as *transmissometers* can provide automatic measurements of visibility. In most types of transmissometer, an optical source produces an intense parallel beam of modulated light. This traverses an atmospheric path, and is received some distance away by a detector, tuned to detect the modulated light in order to reject background light variations.[iii] Mirrors may be used to reflect the beam several times through the same volume of atmosphere, to maximise the optical path length whilst restricting the physical size of the instrument. The detector measures the reduction in beam intensity due to scattering, effectively using Equation 10.2, with x, the total path length of the instrument. Thus ξ can be calculated and converted to an estimate of V, using Equation 10.6. Typical path lengths for transmissometers vary between about 5 m and 50 m, and the maximum visibility they can report is up to about 10 km, less than that of forward scatter meters.

10.2.3 *Present weather sensors*

The optical path and sensors used for a transmissometer can also be used to detect the presence of precipitation in its various forms, and to discriminate between ice, water and snow. A device combining visual range and precipitation discrimination is known as a present weather sensor, as it allows some classification of the present weather conditions. Figure 10.2 shows such an instrument, which uses one infrared beam to determine visual range, and the scattering out of the beam to distinguish between types of hydrometeors if they are present. Precipitation intensity may also be reported.

[iii] Retrieving a modulated weak signal in noise, by its comparison with the original modulating signal is known as *phase sensitive detection*, and has widespread applications. It effectively narrows the bandwidth of the signal, reducing the noise content.

Figure 10.2 A combined visual range and present weather sensor. The left to right infrared beam is used to determine the visual range, and the third sensor used to discriminate between droplets, ice and snow.

10.3 Cloud base measurements

The height of the cloud base is an important visibility parameter for aircraft, and other than manual estimation, can be determined using instruments known as *ceilometers*. The simplest approach for use at night is a cloud searchlight which is arranged to shine vertically upwards, illuminating the cloud base. At a known distance from the searchlight, the angle of elevation to the searchlight beam's position on the cloud base is measured, using an alidade.[iv] The cloud base height can then be obtained by trigonometry (Figure 10.3), with repeated readings made to minimise the uncertainty in the measured elevation (see Section 2.3.2).

Daylight methods of cloud base height measurement require the light reflected or scattered by the cloud to be identifiably different from sunlight, which means that signal processing methods to identify a weak signal in the presence of a strong background signal are again required. A development from the cloud searchlight was the optical cloud base recorder (CBR). This modulated the light beam by a rotating shutter, with the same modulation pattern required in the light received from the cloud. In a UK Met Office system based on this principle, the geometry differed from the searchlight method, in that the optical receiver was vertically beneath the cloud, and the modulated light source displaced away horizontally. The elevation of the beam from the light source was varied until the optical detector receives light from the cloud vertically above. Knowledge of the baseline length between transmitter and receiver and the elevation angle allowed the cloud base height to be calculated. The Met Office Mark 3 version showed an accuracy of 5 to 10%, worse with lower clouds, with a systematic tendency to underestimate the cloud base height [94].

[iv] An alidade is a sighting device which also allows measurement of angle.

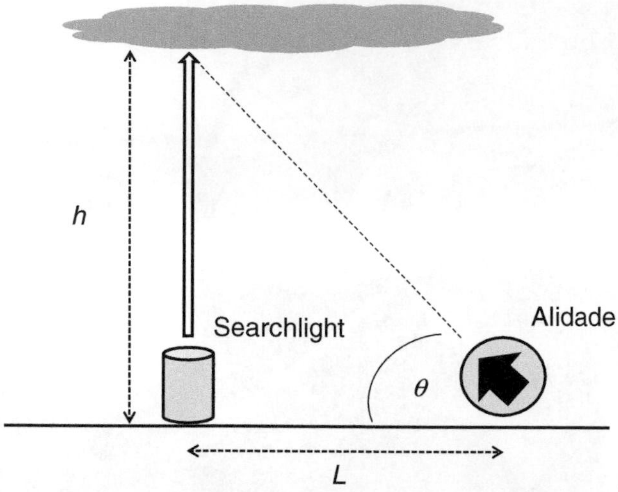

Figure 10.3 Determination of a nocturnal cloud base height h, using a cloud searchlight. By measuring the angle of elevation θ at a distance L from the searchlight, $h = L \tan \theta$.

A more modern approach is the laser cloud base recorder (LCBR), which uses a vertical beam, and is essentially a simple lidar in that a pulse of light is emitted and the returned beam analysed for information. Through timing the interval between an emitted and returned pulse of light, the cloud base height can be calculated. The returns are much stronger for water droplets than for aerosol, allowing the cloud base to be distinguished from other particles present in the lower atmosphere, or calibrated to a certain visual range regarded as representative of cloud.

For all the instrument methods, a major uncertainty in cloud base determination arises from the ambiguity over what constitutes cloud base, and the contribution from precipitation [95]. This is apparent in the difference between measurements obtained using the vertically operating LCBR and the slanting beam of the CBR, which intercept different areas of reflecting raindrop surface.

10.4 Rain gauges

As evident from Figure 1.3, manual determination of daily rainfall amount at a point can be maintained for long periods, through the disciplined use of a collecting vessel of known aperture, monitored and emptied regularly.[v] However, a continuous record of rainfall or the rainfall rate is far preferable, such as for comparison with spatial measurements from satellites or cloud radar. A recording rain gauge is an instrument needing a wide dynamic range, able to remain dormant for long periods of no rain, unattended, but also able to operate reliably during brief occasions of intense rainfall. The need to empty the water collector rapidly whilst continuing

[v] Rainfall amounts are, in the United Kingdom at least, now recorded as depths of rain in mm, which is independent of the area considered. Older records used inches for this; other approaches are to quote an equivalent volume of water per unit area (e.g. 25.4 litres m^{-2} is equivalent to a 1 inch depth of rain).

Figure 10.4 Tilting siphon rain gauge. The rain falls into the float chamber in the centre, and the position of the float determines the position of the pen on a recording chart, normally changed daily. When full, the chamber is emptied by a siphon action.

to measure the rain arriving is a fundamental problem to be addressed in designing a rain gauge.

10.4.1 Tilting siphon

The tilting siphon rain gauge (see Figure 10.4) provides a chart record of rainfall, i.e. it is an autographic instrument. It operates using a pen connected to a float, which moves within a collecting chamber. When the chamber is sufficiently full, it tilts, permitting a siphon to empty it. The chart trace produced has a characteristic 'sawtooth' waveform, caused by the tilting action. The rainfall rate can be determined from the gradient of the rising part of the waveform, or the number of steps in the trace. During siphoning rainfall continues to enter through the collecting funnel, a feature which minimises any 'dead time'.

10.4.2 Tipping bucket

The tipping bucket rain gauge collects rainwater in a funnel and passes it to a pair of small identical buckets balanced on a yoke able to pivot (see Figure 10.5). One bucket fills with rainwater, and when full, the yoke pivots to one side, emptying the first bucket. The second bucket then is then able to fill, and when it is full, pivoting occurs again, in the opposite direction.

Each time pivoting occurs, a pulse is generated by a magnetic or optical switch, which represents a unit of rainfall related to the volume of the bucket. The accumulated number of pulses is proportional to the rainfall amount. If each bucket

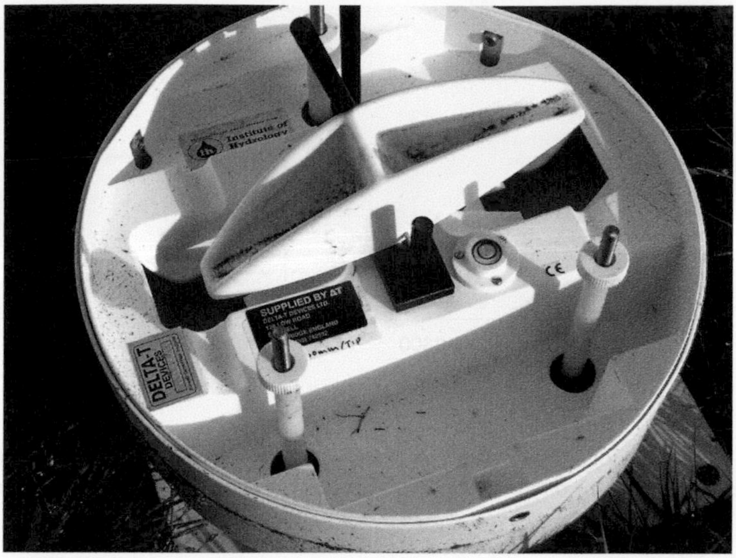

Figure 10.5 Tipping bucket mechanism within a rain gauge. The rain is collected in a funnel and falls onto the uppermost bucket of the two mounted on the sea-saw mechanism. When full, the mechanism tips, emptying the bucket and allowing the other bucket to fill. Each trip action operates a switch, generating a pulse.

fills to an equivalent depth of rainfall B per bucket tip, after N tips the total rainfall P will be

$$P = NB, \tag{10.7}$$

where B is typically 0.5 mm, 0.2 mm or even 0.1 mm. Sensitive tipping bucket rain gauges require small buckets with carefully balanced pivot assemblies.

The generation of tip events represents a point process. If the times at which individual tipping pulses occur are recorded, the rainfall rate can be found from the rate of generation of the pulses. For a tipping bucket device generating pulse i and $(i+1)$ in a series of tips, the instantaneous rainfall rate R is given by

$$R = \frac{B}{t_{i+1} - t_i}, \tag{10.8}$$

where t_{i+1} and t_i are the times at which pulses $(i+1)$ and i occur respectively, and B is the equivalent amount of rainfall per bucket tip. Figure 10.6 shows a typical annual histogram of tip intervals and the derived probability distribution of rainfall rates.

The instantaneous rainfall rate can be attributed to the central time between each pair of pulses, hence the rainfall rate is effectively determined at

$$t = \tfrac{1}{2}\left(t_{i+1} + t_i\right). \tag{10.9}$$

As large rainfall rates generate rapid tips, and with them, more frequent samples of rainfall rate, the temporal resolution in determination of increases with the rainfall rate (see Figure 10.7). The tipping bucket rain gauge is therefore somewhat unusual

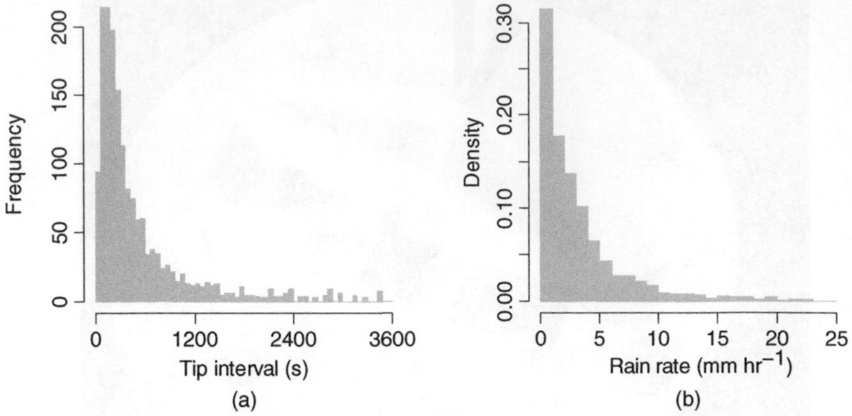

Figure 10.6 Histograms for a year of data from a tipping bucket rain gauge at Reading for (a) intervals between successive tips and (b) the derived rainfall rates. (Tip intervals greater than 1 hour have been removed.)

as an instrument in that temporal resolution in the quantity measured varies with the quantity itself. Its disadvantages are that it does not work with solid precipitation, and that it is susceptible to partial or complete blockage. Neither problem may be evident from the data series alone.

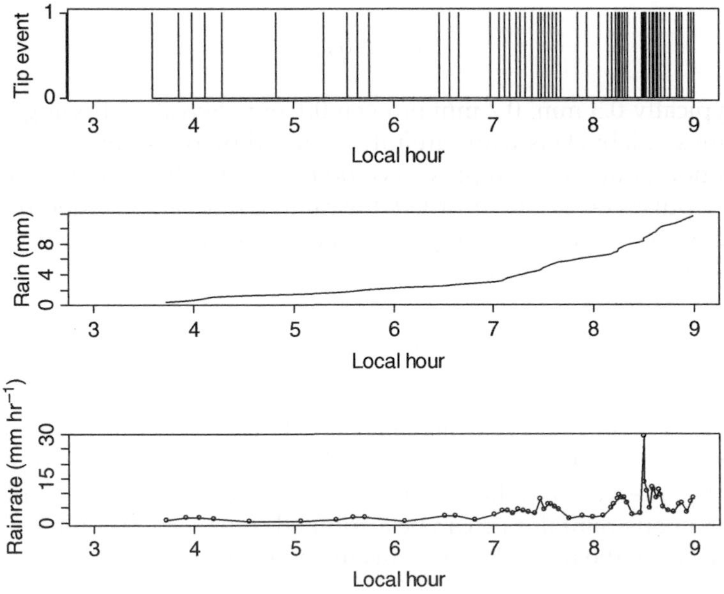

Figure 10.7 Measurements and quantities derived from a tipping bucket rain gauge. The upper panel shows the "barcode" progression of individual tip events, which become increasingly rapid. The middle panel shows the accumulated rainfall, and the lower panel shows derived rainfall rates, with the points indicating the time to which the rainfall rate is attributed.

10.4.3 *Disdrometers*

The rain drop size distribution can be obtained using a disdrometer using acoustic, microwave or laser methods. In the acoustic implementation, a disdrometer operates by the transfer of a falling drop's momentum to a collecting surface or membrane, the motion of which can be detected and is proportional to raindrop size. In essence this is a form of acoustic detection of each droplet as it lands, using a sensor of the moving coil or piezo electric type, similar to that used in audio applications.

10.5 Atmospheric electricity

The atmosphere is always electrified to a greater or lesser extent and it has long been recognised that changes in atmospheric electricity are closely related to meteorological changes.[vi] The most commonly observed quantity in atmospheric electricity is the vertical electric field, which is measured at the surface as the *Potential Gradient* (PG), as this shows strong sensitivity to convectively disturbed weather, lightning and fog formation, and therefore is of value in distinguishing present weather conditions. Conceptually, the PG near the surface is the voltage difference between the surface and a point 1 m above it in electrical equilibrium with the air around it.[vii] The PG arises from currents flowing from distant disturbed weather and thunderstorms globally, coupled through the ionosphere and the planet's surface to fair weather regions elsewhere.

10.5.1 *Potential Gradient instrumentation*

The Potential Gradient (PG) can be measured using a potential probe (also known as a collector) at a fixed height above the surface, as described for fair weather conditions using a long wire in Section 3.3.1. A 'Kelvin water dropper equaliser' is another interesting approach still in operational use,[viii] which uses a fine mist of water droplets from an insulated tank to exchange charge with the local air [39]. A more convenient method for continuous measurements in all weather conditions is an electrostatic instrument or field mill as shown in Figure 10.8. The field mill consists of an approximately circular vaned plate electrode, in which a charge is induced by the atmospheric electric field. The sensing electrode is driven by a motor so that it rotates under an outer mechanical shutter. The induced voltage is averaged, using signal processing electronics which is locked to the rotation rate of the motor by an optical detector.

Field mills generally offer a rapid time response and dynamic range. By ensuring large clearances between the electrode and shutter, field mill instruments can be made sufficiently robust to allow operation in rain, hail, and snow. The atmospheric field

---

[vi] Lord Kelvin famously remarked at the Royal Institution in 1860 that '... there can be no doubt but the electric indications, when sufficiently studied, will be found important additions to our means for prognosticating the weather.'

[vii] Note that this has the opposite sign to the conventional electrostatic definition of electric field.

[viii] A water dropper is still used at Kakioka Observatory in Japan; it detected PG changes associated with the radioactivity released from the Fukushima reactor accident.

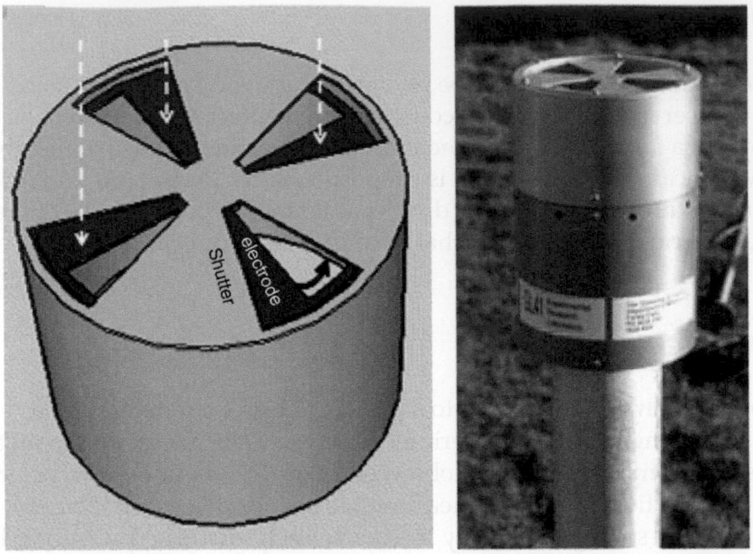

Figure 10.8 (Left) Schematic of an electrostatic field mill. A vertical electric field (dashed lines) induces charge on a rotating sensing electrode, alternately exposed and concealed under a shutter. (Right) A JCI131 commercial field mill installed at Reading (diameter ~ 10 cm). (Reproduced with permission of John Wiley & Sons.)

is, however, itself distorted by the presence of the field mill and its support mast, so a correction is needed for this to provide accurate values. This is usually obtained during fair weather conditions by using a long horizontal wire antenna nearby for calibration, as the long wire collector has negligible associated distortion and its supports contribute little error to the PG obtained (see also Section 3.3.1).

10.5.2 *Variability in the Potential Gradient*

In fair weather conditions and unpolluted air, the PG at the surface is between about 100 V m^{-1} and 150 V m^{-1}. If the fair weather values are averaged for the same times each day, a variation with a single maximum at 19UT and minimum at 03UT is seen, known as the *Carnegie* curve (see also Section 12.4.3). This global variation is almost never seen on a single day, because the PG variation is more strongly affected by local factors, such as abrupt changes in weather (see Figure 10.9).

The PG is closely related to the electrical conductivity of air, the charge on passing clouds and the charge carried by precipitation. Fogs, clouds and precipitation can therefore cause marked changes in the PG. During fogs, in which the water droplets scavenge the ions responsible for air's electrical conductivity, the magnitude of the PG substantially increases. When there is liquid precipitation, the PG changes by ~500 V m^{-1} or greater, commonly with a reversal of sign to negative PG. This is due to charge contained within the cloud, and that transported by falling raindrops. In thunderstorms, the surface PG can exceed ±10 kV m^{-1} before a lightning discharge, and usually becomes very variable. Figure 10.10 summarises typical PG changes associated with disturbed and fair weather conditions.

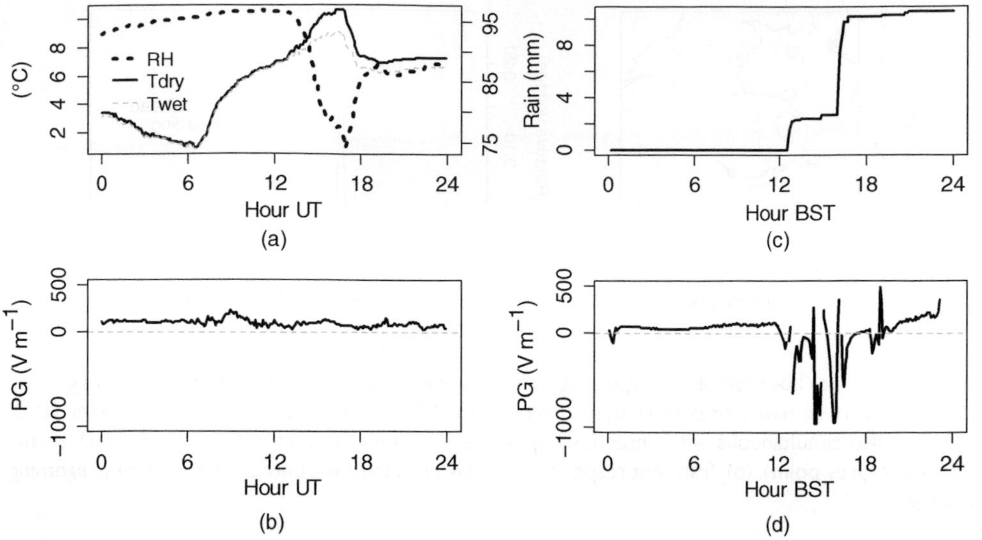

Figure 10.9 (a) and (b) Variations in humidity variables and Potential Gradient (PG) during morning fog on 2 March 2012, after which clearer, fair weather conditions became established (c) and (d) show rainfall and PG changes respectively around thundery rain during the afternoon of 7 August 2011. The gaps in (d) arise from the large transients associated with lightning.

10.5.3 Lightning detection

Transient changes in the PG provide evidence of local lightning discharges within or between clouds, or between cloud and ground. This electrostatic method of lightning

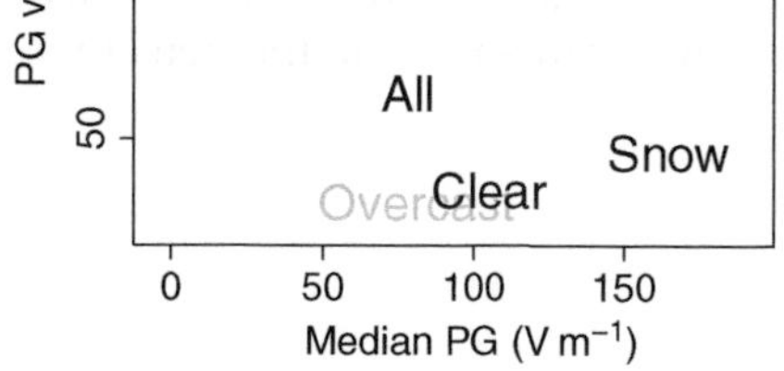

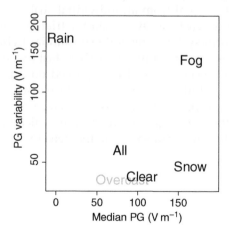

Figure 10.10 Summary of typical values of the Potential Gradient (PG) measured at Reading in different weather conditions, together with an estimate of its associated variability calculated from the interquartile range. Criteria used to identify weather conditions: **Clear** – Diffuse Fraction (DF) ≤ 0.4; **Overcast** – DF ≥ 0.9 and no rain; **Rain** – rainfall > 0 mm; **Fog** – equal dry bulb and wet bulb temperatures or Relative Humidity $\geq 90\%$; **Snow** – manual identification. (Modified from [96]).

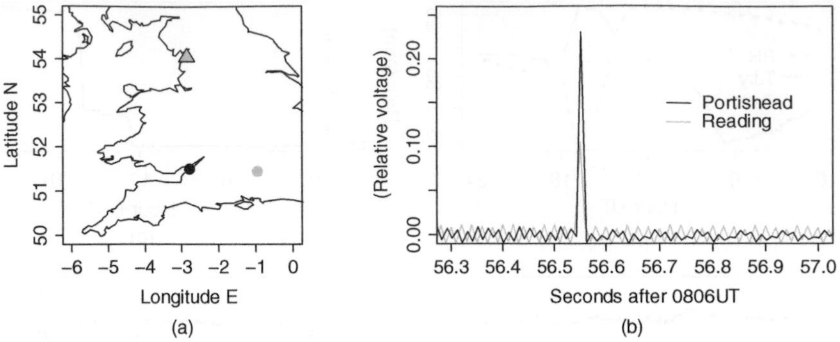

Figure 10.11 (a) Location of a UK lightning strike (triangle) determined by a radio frequency lightning location network to have a peak discharge current of 267 kA, just after 0806UT on 2 November 2012, which was also simultaneous with detections by two electrostatic sensors at Portishead (black point) and Reading (grey point). (b) Transient response of the two electrostatic detectors to the same lightning discharge.

detection has limited range, as the detection of transients due to lightning falls off with the cube of the distance between the detector and the lightning. If there are large charge structures associated with a storm such as halo-like sheets of charge above the cloud, detection may be possible at greater distances [95].

Radio methods for lightning detection have greater operating range, and mostly operate at frequencies in the range of 10 to 100 kHz, where the electromagnetic energy radiated by a lightning strike is relatively strong. If a directional antenna is used on the receiver, the signal strength can be used to provide an estimate of the distance to the lightning discharge. A more sophisticated and accurate method is to use a combination of distributed receivers, all of which are able to rapidly digitise the waveform generated by the lightning strike. By identifying common waveform shapes at the multiple receivers, the signal from an individual strike can be uniquely identified, and the bearings from each receiver used for triangulation to find the source region. Because some storm systems generate substantial lightning flash rates, and some lightning discharges consist of multiple strikes, fast processing is needed. Optical methods of lightning detection are also employed on satellites, but, because of range limitations, have seen little use in routine surface instrumentation.

Figure 10.11 shows the response in electrostatic detectors at two locations at considerable distance from the same lightning discharge, which vary in amplitude with the distance from the lightning event source to the detector.

11
Upper Air Instruments

Measurements made above the surface are important for forecasting analyses and for research applications because of the information they provide on the atmosphere's vertical structure. Data obtained within the atmosphere (i.e. *in situ* measurements) can be provided by sensors carried on aircraft or balloon platforms. Balloon-carried sensors are routinely used internationally to provide data for forecast initialisation, launched at thousands of sites daily as well as more specialist surveys, for example to monitor stratospheric ozone. In some cases, such as emergency situations concerning the release of radioactivity or volcanic eruptions, balloon carried probes can provide the only source of safe *in situ* sampling throughout the troposphere. A balloon-carried measuring probe returning data by radio is known as a *radiosonde*.

11.1 Radiosondes

Radiosondes instrument packages are carried by balloons to the upper troposphere or above, during which vertical profiles are obtained of pressure, temperature and humidity. In addition, changes in position during the ascent are used to derive the wind speed and direction. Measurements are telemetered to a ground station by a radio transmitter [97]. Radiosondes were developed from mechanical recording devices carried by balloons, with the first meteorological radiosonde measurements made during the early 1930s. Since then, many radiosonde devices have been constructed internationally. Some independent designs are still produced by individual national meteorological services, but there are several commercial manufacturers who supply their own design of radiosonde in quantity, to sustain the large number launched daily. A major European supplier of radiosondes is the Finnish company Vaisala,[i] but several other commercial radiosonde packages are available, notably from Meteomodem (France), Graw (Germany), Meisei (Japan), Sippican (United States), as well as from India, South Korea and China.

[i] This company was founded by Vilho Väisälä, (1889–1969), a meteorologist and physicist, whose first commercial radiosonde, the RS11, was completed in 1936. The current Vaisala radiosonde is the RS92, which uses digital technology and data transmission. Its analogue predecessor, the RS80, is still used.

Meteorological Measurements and Instrumentation, First Edition. R. Giles Harrison.
© 2015 John Wiley & Sons, Ltd. Published 2015 by John Wiley & Sons, Ltd.
Companion website: www.wiley.com/go/harrison/meteorologicalinstruments

11.1.1 *Sounding balloons*

Beyond the sensor package itself, the carrier balloon is fundamental to the atmospheric sounding capability of the radiosonde. For this, a latex balloon of mass between about 200 g and 1000 g is inflated with the lighter than air gases of helium or hydrogen to allow free lift to be obtained. Of these gases, helium is becoming an increasingly scarce resource and expensive; hydrogen is less costly as a raw material, but, because it forms an explosive mixture with air, it requires special handling to prevent the possibility of ignition. The balloon eventually bursts from expansion, and the radiosonde package descends by parachute. Although measurements can be made on the descent if there is battery capacity remaining, these are generally not used in operational meteorology because of the different sensor environment encountered and the displacement from the launch site. Depending on the circumstances the instrument package may be able to be recovered, but it is more commonly regarded as disposable. The flight train of balloon, parachute and radiosonde fixing is shown in Figure 11.1.

The altitude reached by the balloon when it bursts depends on the weight of its payload, the drag forces and the duration of the flight, as well as whether there are imperfections in the latex from which it is made. Because the shape of the balloon changes during its ascent, the drag force is not easily calculated and empirical relationships have to be used. Figure 11.2 shows the typical variation in burst height with balloon mass. Larger balloons reach greater burst heights for the same payload: compared with the range of burst heights, there is much less variation in ascent speed with the balloon size.

Figure 11.1 Sounding balloon just after launch, carrying a radiosonde instrument package (out of view). Beneath the balloon is a parachute, used to control the descent after the balloon bursts.

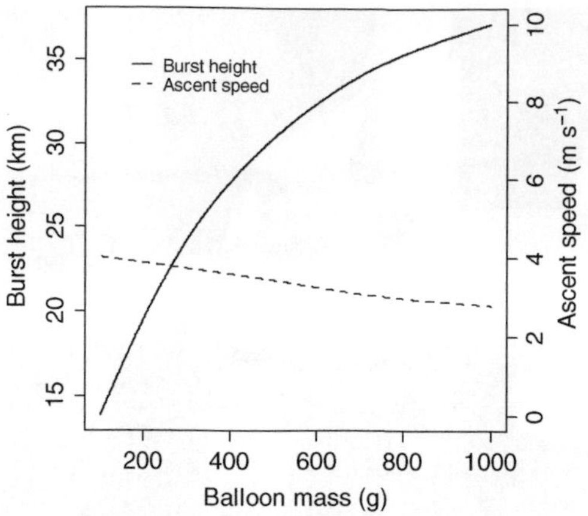

Figure 11.2 Variation of flight parameters with the unfilled mass of the carrying balloon, calculated for Helium as the lifting gas, with a payload of 365 g.

11.2 Radiosonde technology

Because radio telemetry is used to return the atmospheric data, the sensors carried by a radiosonde have to provide an electrical output, such as a voltage, current or frequency. The earlier radiosonde technologies provide a good basis with which to understand the modern operating principles, as they employed simple sensor techniques and straightforward methods of sending the information back by radio. Following the early Vaisala radisonde of the 1930s, the UK Met Office produced several of its own design of radiosonde [98], until discontinuing its final (Mark III) version in the 1970s; a Mark II Met Office radiosonde, the design used between the 1940s and 1960s, is shown in Figure 11.3.

In the Mark II Met Office radiosonde, each sensor was mechanical, generating a change in dimension with change in pressure (aneroid capsule), temperature (bimetallic strip) or humidity (goldbeater's skin).[ii] These mechanical changes were used to vary the properties of an inductor, and, in turn, modulate the frequency of an audio oscillator. The audio oscillation was sent via the radio transmitter and antenna system, usually a long wire trailing beneath the radiosonde. A summary schematic is given in Figure 11.4.

A key requirement for a radiosonde is its battery power supply, which must sustain the operation of the device as it becomes colder and more distant from the receiving station. Early radiosondes used lightweight versions of 'wet' (lead-acid) batteries, but these are now rarely used as power sources employing primary cells such as alkaline batteries or even lithium batteries have become commonplace. The reduction in battery capacity with temperature means that the power consumed by the radiosonde electronics is a consideration in determining the mass of the battery which must be carried. This requires a compromise between transmitter power (and therefore signal range), and battery capacity (and therefore operating duration). Efficient radio systems, with sensitive receivers and high-gain or tracking antennas can also reduce the transmitter power needed to obtain a substantial radio range.

[ii] The combined measurement of pressure (*P*), temperature (*T*) and relative humidity (*U*) is often abbreviated to 'PTU'.

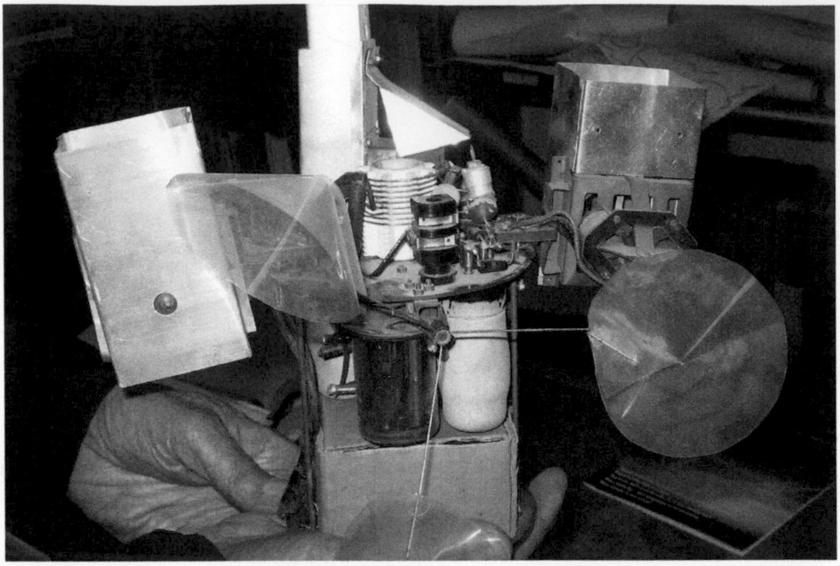

Figure 11.3 Early Mark II Met Office radiosonde, showing (centre) the thermionic valves used to interface the mechanical PTU sensors and generate the radio signals. The three sensors are housed in the metal radiation shields, allowing air to flow vertically. On the right-hand side, beneath a sensor, the hexagonal sensing inductor can be seen. In the foreground are the mesh vanes of a small windmill, which rotated as the radiosonde ascended. The battery was mounted in the base of the device.

Requirements for modern radiosonde sensors are demanding, as they are expected to measure consistently and accurately across a wide range of conditions. This might typically include pressure from ~1000 hPa to ~20 hPa to ±1 hPa, temperature from ~40°C to ~ −60°C to ±0.5°C and relative humidity from ~0 % to ~100 % to ±5%. A further complication is that, if a radiosonde passes through a cloud, the

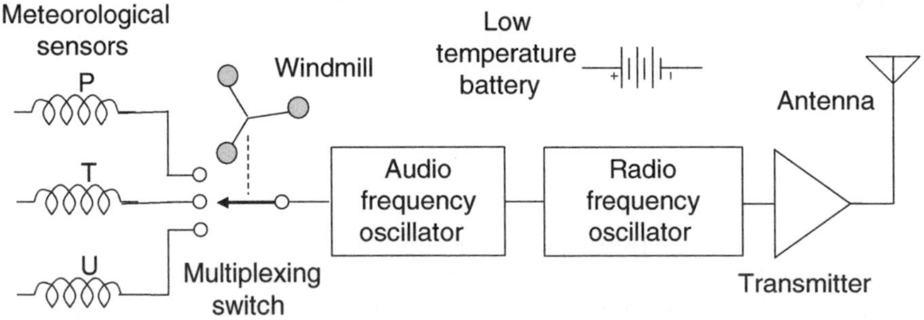

Figure 11.4 Operating principle of the radiosonde shown in Figure 11.3. Each of the mechanical pressure, temperature and humidity sensors caused a change in an inductance, which controlled the frequency of an audio oscillator. The different inductors were switched in turn to determine the audio frequency oscillator, by the mechanical rotation of the windmill. These distinct, sequential audio tones were used to modulate a radio frequency oscillator, amplified by a transmitter stage to radiate from an antenna. An efficient battery able to operate at low temperatures is also needed.

temperature sensor may become wet and provide measurements of wet bulb temperature as it emerges from the cloud, or even become coated in ice. Radiosondes may also encounter vigorous turbulent motion associated with rapid updrafts or jet streams. In addition, the full radiosonde package needs to be lightweight to minimise the amount of lifting gas required, as well as being disposable and inexpensive.

11.2.1 *Pressure sensor*

Pressure is usually measured on a radiosonde using an aneroid capsule, with temperature compensation for the wide thermal variations encountered. The transducer converts the small movement of the capsule into a capacitance or inductance variation (see Figure 7.4), or may operate a switch, with different positions corresponding to different heights. The pressure sensor will be mounted within the radiosonde, usually also within some lightweight thermal insulation, such as polystyrene. Figure 11.5 shows examples of pressure sensing capsules.

An alternative method of determining pressure is to use the boiling temperature of a liquid, which varies with pressure (see Section 7.2.3). Such a hypsometer becomes more sensitive at low pressures, when the aneroid capsule may be less sensitive. An important advantage over an aneroid capsule is that only a temperature measurement is required, without the complexity of the corrections needed for an aneroid with its changing sensitivity. To avoid the power required in heating, the sensing liquid used can be chosen to have a boiling point less than that of the ambient temperatures expected to be encountered (e.g. by using a chlorofluorocarbon). Alternatively, a very small sensing cell requiring little power can be used, to allow the use of water and minimise the environmental hazard. For a water hypsometer (see Figure 11.6) giving

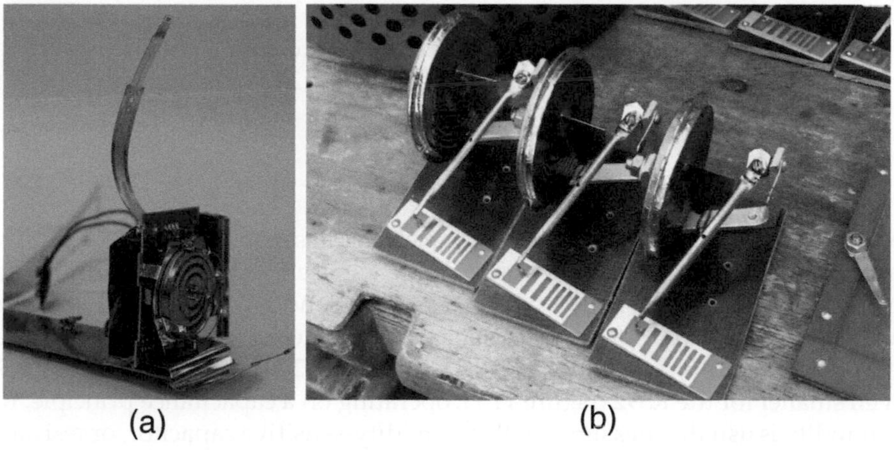

Figure 11.5 Pressure sensors for balloon instruments. (a) View of a pressure sensor chamber within a dismantled Vaisala RS80 radiosonde. (The vertical strip above it carries the humidity and temperature sensors, which protrude beyond the final cardboard and polystyrene housing.) (b) Pressure-switches (baroswitches) manufactured by the Lebedev Institute, Moscow, to determine fixed heights reached by their cosmic ray radiosondes. The irregular spacing chosen for the baroswitch contacts compensates for the non-linear relationship between pressure and height.

Figure 11.6 Hypsometer element (centre) used within a radiosonde. Heating is provided by small resistive heating elements, and the boiling temperature determined with thermocouples. (The foil wrapped heater is about 3 cm long.)

pressure measurements to 0.5 hPa, temperature measurement to 0.01°C is required, which can be achieved using thermocouples [99].

11.2.2 Temperature and humidity sensors

Air temperature is usually measured using an electrical sensor, optimised for a small exposed area to minimise radiation errors. A fine wire sensor has been used on some designs of radiosonde (e.g. the Met Office Mark III sonde, which used fine (13 μm diameter) tungsten wire to form a resistance thermometer), but the recent Vaisala sondes use a temperature-sensing bead or wire, about 2 mm diameter for the RS80 and even smaller for the RS92 (Figure 11.7), operating on a capacitance principle. Relative humidity is usually measured with a humidity-sensitive capacitor, or resistance sensor (hygristor). In the Vaisala RS92, a pair of capacitance RH sensors is used (see Figure 11.8); these sensors alternate between measuring and being heated to avoid wet bulb saturation effects (see section 11.3.3).

The Vaisala RS80 and RS92 sensors are factory-calibrated using a polynomial function to characterise the individual sensor responses, which is unique to each device and applied at the receiver in the data processing. Initial values of station pressure,

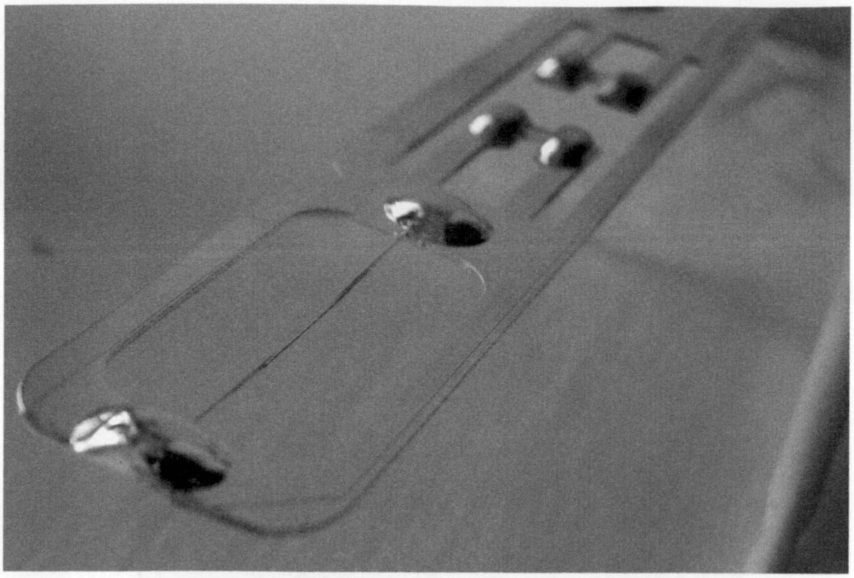

Figure 11.7 View along the sensor strip of a Vaisala RS92 radisonde. In the foreground is the temperature sensing wire element (length 20 mm between connections), and, further back, there are two relative humidity sensors, which are capacitive elements mounted between connections. The whole assembly is silvered to minimise radiation errors.

temperature and humidity calibration are still made as checks before release, and the RS92's pair of RH sensors are matched at the surface before release in a small calibration chamber, which is dried using a desiccant.

11.2.3 *Wind measurements from position information*

Early radiosondes used radar tracking to derive upper-air wind velocities. They carried a radar reflector formed from metallised mesh, which reflected radio waves with high efficiency. The tracking radar measured the time of travel of a radio pulse from radar antenna to sonde and back to antenna to determine the range, together with the inclination above the horizontal, and the bearing. From successive measurements of position, the three-dimensional velocity vector was calculated, as well as the altitude reached by the radiosonde. Typical accuracy using the tracking approach was about ± 1 m s^{-1} at a height of 7 km, with a height accuracy of about ± 40 m.

Radio navigation systems are now widely available which give an exact location in space and time. In the LORAN system, very low frequency radio signals are transmitted synchronously from radio transmitters at widely separated locations. Using LORAN, a radiosonde receives signals from different transmitters, and the time difference in their reception measured to determine its position. Alternatively, Global Positioning System satellites are now almost always available. Several GPS satellite transmitters are received by the radiosonde, each of which sends a unique identifier allowing the position to be determined. Wind speeds above the surface can be

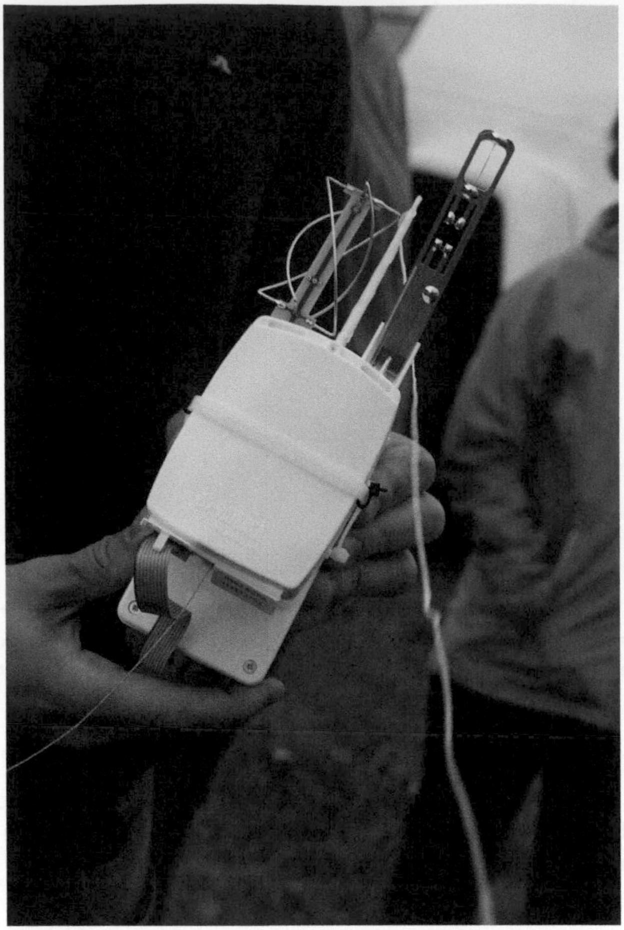

Figure 11.8 A Vaisala RS92 radiosonde, immediately prior to launch. The sensor strip carrying the RH and temperature sensors is uppermost on the right-hand side of the device, and the helical GPS receiving antenna is on the left-hand side. At the bottom of the package curved to the left is the UHF radio antenna. The string for attachment to the balloon is trailing into the foreground. (This particular radiosonde is also carrying a "Pandora" box equipped with additional science sensors, shown in more detail in figure 4.8.)

determined from differences in successive positions of the radiosonde, to about $\pm 2\,\mathrm{m\,s^{-1}}$. The power required for the GPS direction finding receiver carried can sometimes be a consideration in an instrument's operating duration.

11.2.4 Data telemetry

Changes in the electrical outputs from the radiosonde transducers can be conveyed to the radio transmitter in different ways. As Figure 11.4 shows, one approach is to arrange for an oscillator to vary with the parameter sensed, and then supply the oscillation frequency to modulate the radiosonde transmitter directly. If there is more than one sensor then switching will be needed, which is a simple example of time-division multiplexing, allowing sequential transmission of multiple signals through a single data channel (see also figure 4.4). Multiplexing is well-suited to this application, as variations in the meteorological parameters sensed occur less rapidly than a channel switching rate of about one channel per second, as used in the Vaisala RS80 analogue radiosonde. By ensuring that the range of frequencies associated with each sensor cannot overlap with another, and knowing the order in which the signals are sent, the individual information from each sensor can be recovered. The repeated switching between different audio tones provides a distinctive sequence of sounds to the ear, within which a slow change in pitch can be identified during an ascent or descent.

The most modern radiosondes, such as the Vasiala RS92, do not use variable audio frequency telemetry, but use digital data transmission instead. For this, the radiosonde includes analogue to digital converters, to digitise the measurements from the sensors and a microcomputer to merge and control the data streams from the multiple sources (Figure 11.9). Once in digital form, the information is sent over the radio channel using a modem, which converts the digital information in the form of bits to two distinct tones which can be identified at the receiver to reconstruct the original digital information.

11.2.5 *Radio transmitter*

A simple radio transmitter provides the data telemetry link to the distant ground station. This is a single channel device, at about 400 MHz in Europe, but other frequencies in the shortwave range are used in other systems as allocated by international radio regulations. The radio frequency signal has to have the measurement information added to it, by some form of modulation. At its simplest, this can be achieved by switching the carrier wave on and off, known as carrier wave modulation, but it can also be achieved by varying the signal strength (amplitude modulation), or varying the carrier wave's frequency (frequency modulation).

The radio frequency for the transmitter (or carrier wave) is generated by an oscillator, which may be derived from a resonant circuit using an inductor and capacitor, or, to reduce effects of thermal drift, a quartz crystal oscillator. Once generated,

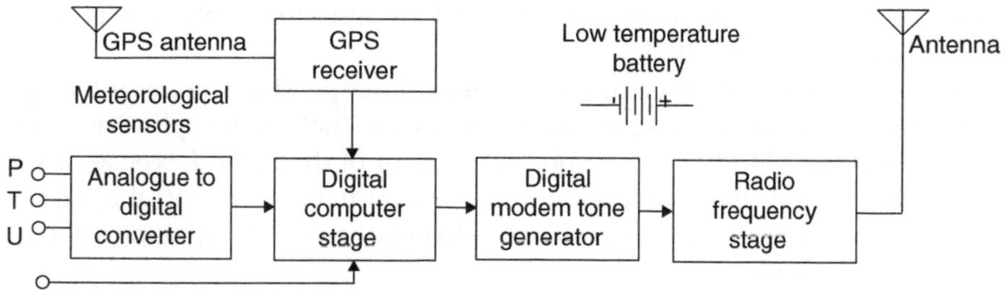

Figure 11.9 Conceptual arrangement of a digital radiosonde. Measurements from the pressure, temperature and relative humidity (PTU) sensors are merged with any other measurement data (shown as an uncommitted connection node) and position data from a GPS receiver by a small digital computer. The data stream generated is passed to a modem, to convert the digital sequence into two different audio tones for radio transmission. These digital tones are converted back to data values at the receiver.

the radio frequency is supplied to an antenna, which can be made resonant to optimise the strength of the signal radiated. Typical line of sight ranges obtained with radiosonde transmitters are 50 to 100 km, but this depends on the terrain, as, if the radiosonde descends below the horizon, the signal will almost always be lost.

11.3 Uncertainties in radiosonde measurements

11.3.1 *Response time*

The response time of a radiosonde sensor determines the vertical resolution at which information can be obtained. For a device rising at about 5 m s^{-1}, response times of 5 to 100 s imply vertical smoothing over about 25 to 500 m of altitude. Temperature sensors generally have a rapid time response, typically a few seconds or less, but capacitance RH sensors have a finite response time which increases as their temperature decreases [100]. Figure 11.10 shows measurements of temperature and humidity from both a radiosonde ascent and its subsequent descent by parachute, when there was little horizontal displacement of the radiosonde. The vertical structure in temperature is similar in both cases at the lower levels, but there is more variability in the RH. Whilst there may be more local variability in moisture field than the temperature field, the broad structure of the RH variation is similar for both ascent and descent, hence some of the variability observed in RH may arise from the time response of the RH sensor. (See also Figure 11.16).

11.3.2 *Radiation errors*

The temperature sensor in a radiosonde may be exposed to full sunshine through most of a daytime flight and, as for a thermometer at the surface, is subject to a radiation error. The radiation error ΔT varies with sensor diameter, its reflectivity and the ascent speed and the reflectivity of the sensor. This increases with increasing height as the incoming solar radiation tends towards its maximum (top-of-atmosphere) value. Differences between designs of radiosondes lead to different temperature errors [101], and radiation corrections are important if long term tropospheric temperature changes are to be examined [102].

An estimate of the radiation error for a spherical temperature sensor intercepting solar radiation can be found using a similar analysis to that for the cylindrical sensor of Section 5.5.1 Figure 11.11 shows a spherical sensor of diameter d exposed to solar radiation from above. If the area intercepting solar radiation is A and the area from which thermal losses occur is A', the balance between the rate of acquisition of energy and the heat loss is given by

$$S(1-\alpha)A = A'\left[\sigma\left(T^4 - T_a^4\right) + \frac{k(T-T_a)}{d}N\right], \tag{11.1}$$

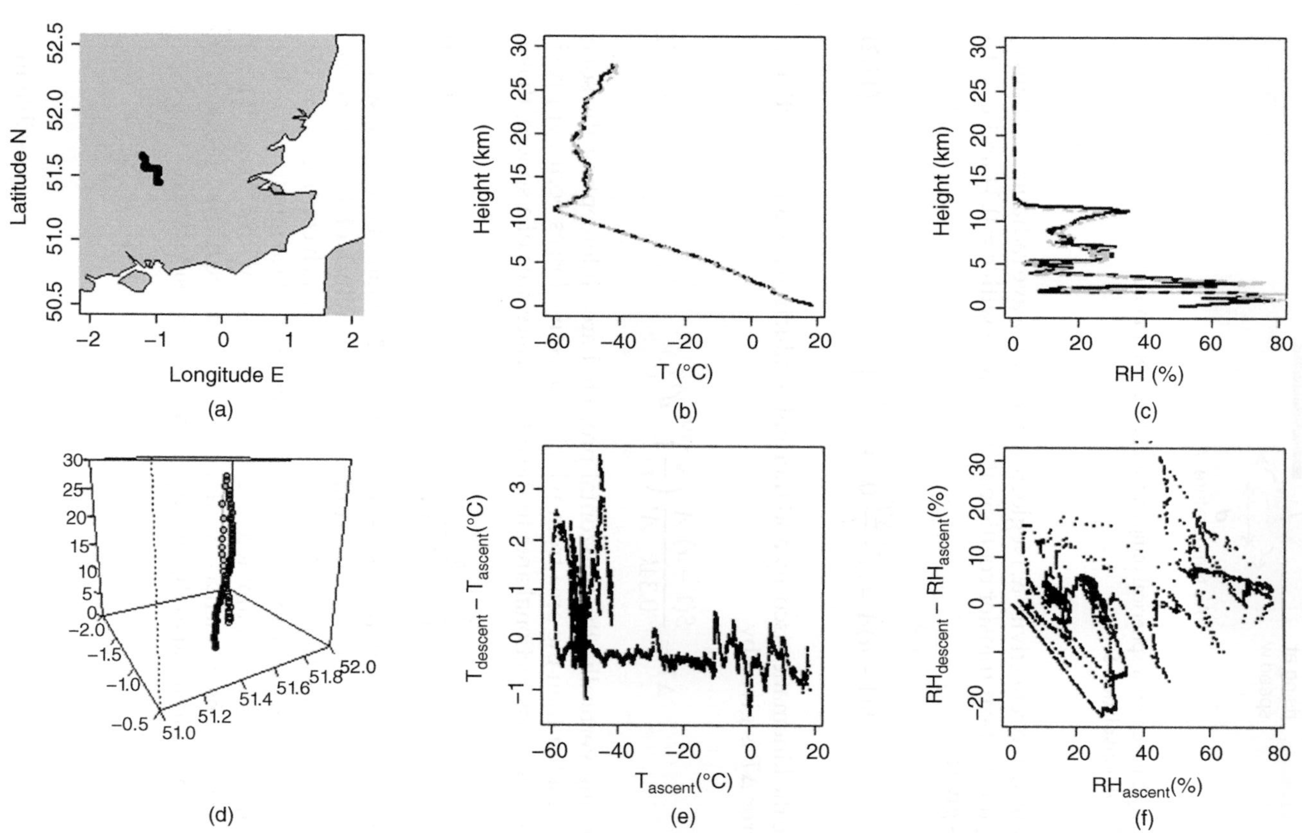

Figure 11.10 Comparison of ascent and descent data for a radiosonde flight using a Vaisala RS92 radiosonde (6 June 2013). (a) Trajectory of entire flight. (b) and (c) soundings of air temperature and relative humidity (*ascent* solid black line, *descent* dashed grey line). (d) Perspective plot of ascent and descent (grey scale on points signifies elapsed time on measurements, from black to grey). (e) and (f) Air temperature and relative humidity differences between ascent and descent interpolated to the same height, plotted against ascent values.

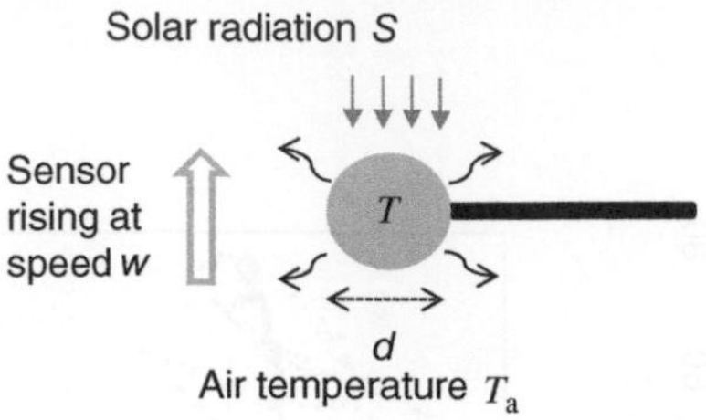

Figure 11.11 Energy balance for a spherical temperature sensor of diameter d rising at a speed w in air of temperature T_a with insolation S.

for a sensor of short wave reflectivity α. Neglecting the long wave radiation exchange, and substituting for the heat transfer coefficient $N = c(Re)^{\zeta}$ with $c = 0.34$ and $\zeta = 0.6$ for a sphere, this gives

$$S(1-\alpha)A = A'\frac{k\Delta T}{d}0.34\left(\frac{wd}{\nu}\right)^{0.6} \quad (11.2)$$

where ν and k are the kinematic viscosity and thermal conductivity of air respectively. The radiation error ΔT is given by

$$\Delta T = \frac{S(1-\alpha)}{0.34k}\frac{A}{A'}\left(\frac{\nu}{w}\right)^{0.6} d^{0.4}. \quad (11.3)$$

For a sensor rising vertically, the horizontal projected area intercepting the solar radiation will be smaller than the area from which convective losses occur, approximately the ratio of the cross sectional area to the total surface area of a sphere, that is

$$\frac{A}{A'} \approx \frac{\pi d^2}{4}\frac{1}{\pi d^2} = \frac{1}{4}. \quad (11.4)$$

Figure 11.12 shows the radiation error estimated using Equation 11.3, for assumed ascent rates, an insolation S of 1000 W m^{-2} and sensor reflectivity α of 0.5. This shows that ΔT is non-linearly proportional to the size of the sensor, which confirms expectations from Section 5.5.1 that temperature sensors should be made as small as possible. In terms of the other controllable parameters, ΔT varies linearly with α, hence if α can be increased for a small sensor, such as by silvering it, ΔT can be further reduced. In practice, the long wave radiation exchange also becomes a consideration.

11.3.3 Wet-bulbing

If a temperature (or humidity) sensor has passed through a cloud, it is possible that the sensor may remain wet for some time after emerging from the cloud. As with a wet bulb thermometer, the local evaporation can lead to additional sensor cooling, and, in turn, degraded accuracy. In the case of air temperature measurements, this effect can be important, particularly in identifying the cloud air boundary.

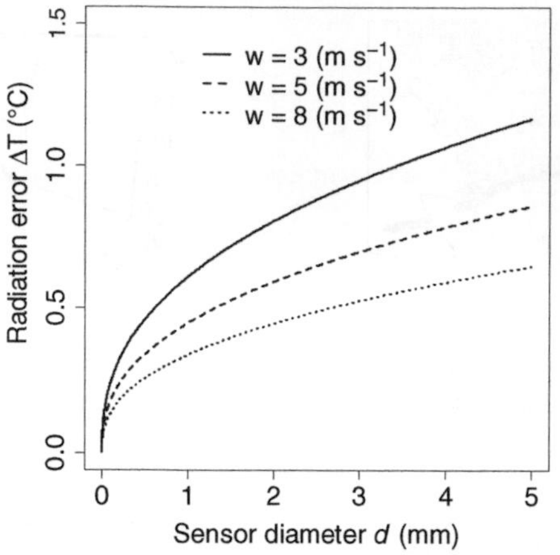

Figure 11.12 Variation of radiation error ΔT for spherical temperature sensor with sensor diameter d, for assumed ascent rates of 3 m s^{-1}, 5 m s^{-1} and 8 m s^{-1}. (Assumptions: Insolation $S = 1000$ W m^{-2}, kinematic viscosity of air $\nu = 1.5 \times 10^{-5}$ m^{-2} s^{-1}, thermal conductivity of air $k = 2.5 \times 10^{-2}$ W m^{-1} K^{-1}, ratio of radiation interception area to convective loss area 0.25, sensor reflectivity $\alpha = 0.5$.)

11.3.4 *Location error*

It is frequently assumed that a radiosonde rises vertically, and that the data obtained therefore represents an atmospheric sounding directly above the launch site, but this is generally not the case and indeed the radiosonde's displacement is used to derive the wind speed and direction as the device ascends. As Figure 11.13 shows, some radiosondes can cover considerable distances.

Figure 11.14 shows two further examples of trajectories with appreciable horizontal displacements, to illustrate the complex motion which can occur during a sounding as updrafts and downdrafts are encountered. The profile of the meteorological data obtained is effectively characterising the properties of a region, defined by the strength of the horizontal wind speeds encountered. Consequently the upper air data obtained may be appropriate to a location significantly displaced from the launch site.

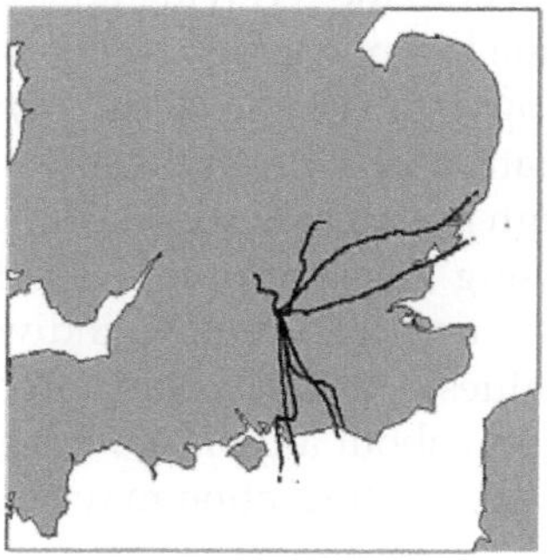

Figure 11.13 Examples of trajectories of radiosondes launched from Reading, up until the telemetry was lost or became intermittent. (The radiosonde position was determined using GPS information and the height calculated from the radiosonde pressure measurement.)

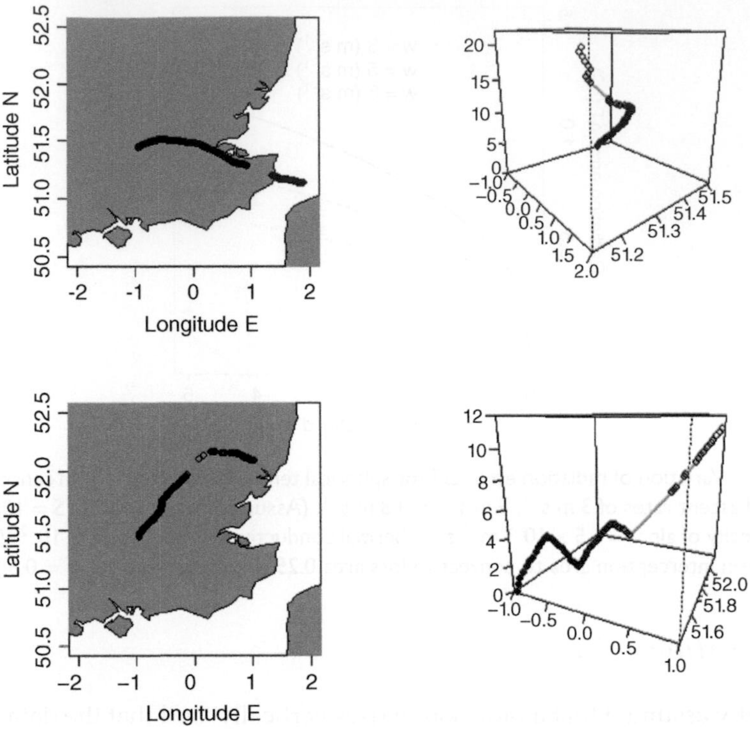

Figure 11.14 Trajectories of radiosonde ascents launched from Reading, shown (left-hand panels) as variation in geographical position and (right-hand panels) as perspective plots of the motion. The vertical axes on the right-hand panels have units of km; the horizontal axes show the longitude east and latitude north.

11.3.5 Telemetry errors

If the radio signal is weak due to the distance travelled between the transmitter and receiver or because of the depleted state of the battery, or poor alignment of the radiosonde's antenna with respect to the receiving antenna, the radio telemetry may become intermittent and the data lost or corrupted. The simplest method of data processing for corrupted or missing data is to delete unphysical values, that is remove those readings which cannot naturally occur in the terrestrial atmosphere and therefore can only have arisen through poor telemetry or a deficient sensor. A more sophisticated method of data processing is to compute a slowly varying moving average from all incoming data values,[iii] and then remove individual values deviating substantially from the smoothed values. This can be very effective for identifying outliers, but it does require an assumption about a typical averaging timescales, and with it, an acceptance that some small-scale information may be lost.

[iii] A moving average is an example of a *low pass filter*, as it retains slow (low frequency) variations in the data, whilst removing the high frequency variations.

11.4 Specialist radiosondes

Beyond the standard PTU meteorological data, radiosondes provide inexpensive measurement platforms which can be used for a variety of atmospheric sensors. Amongst other quantities, they have been used to sense ozone concentrations, radioactivity, cloud electrification, solar radiation and turbulence. They are a particularly useful platform for such measurements because they already include access to power, radio telemetry and position data, and therefore any additional measurements are essentially only an increment on the existing infrastructure. Clearly any additional sensors carried by a radiosonde are also required to be small and lightweight, and to consume little extra power.

A method of sending the extra measurements to the surface, without compromising the existing meteorological and position data is also needed. Some digital radiosondes provide access to the internal data system to allow extra channels to be added (see Figure 11.9 and Section 4.2.2 for an example system) without the need for additional receiver hardware, but for analogue radiosonde systems using variable audio frequencies, superimposing the extra data channels on the existing data telemetry will be needed. One method to achieve this is to use a modem with frequencies for digital transmission which lie outside the range of frequencies used for the analogue sensors, with a receiving modem at the ground station [103]. A disadvantage in general is that data rates available can be limited to a few bytes per second,[iv] requiring either that measurements are obtained fairly slowly, or that some processing of the raw data is required to derive summary statistical parameters before onward transmission.

11.4.1 Cloud electrification

Understanding the electrification of thunderclouds led to an early specialist radiosonde, the alti-electrograph [104], which sensed the corona current generated in an electrode in response to locally intense electric fields. This device was able to determine the relative polarity of nearby charge, and hence deduce the electrical structure of thunderclouds. Subsequent radiosonde measurements made within thunderclouds now carry sensors related to the field mill principle [105]. Electrification of fair weather clouds has also been measured using radiosondes, for example associated with the upper and lower horizontal edges of extensive layer clouds [106, 107].

11.4.2 Ozone

The measurement of atmospheric ozone is important because of its role in shielding harmful ultraviolet radiation and in controlling heating in the stratosphere. Atmospheric ozone is measured electrochemically on radiosondes (known as ozonesondes), using a technique [108] developed in the 1960s. Air is pumped through

[iv] Data compression can provide improved use of the available data bandwidth. Significant intervals in the digital data sent, however, may cause the receiver to shut off or to lose synchronisation. This can be avoided by choice of a transmission protocol, such as *Manchester encoding*, which balances the high and low bits sent.

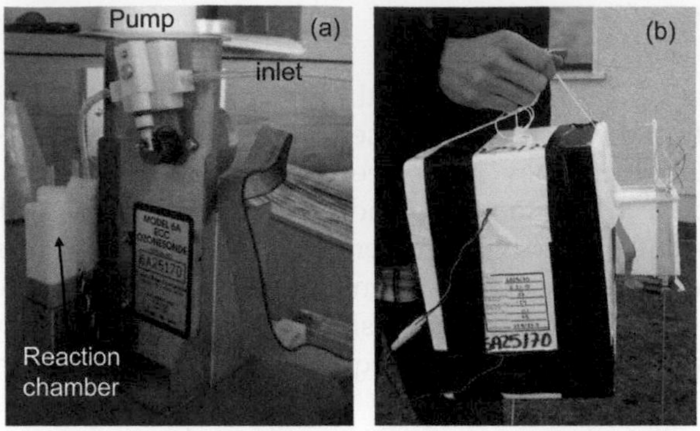

Figure 11.15 (a) View during the preparation of a radiosonde ozone sensor, showing the inlet pipe, pump and reaction chamber. (b) Ozone sensor (left) integrated with an RS92 radiosonde (right).

potassium iodide solution, and the reaction of potassium iodide with ozone releases iodine, allowing a small current to flow. The current is proportional to the ozone concentration. Originally this approach was used with the Mark II Met Office radiosonde, by replacing one of the sensor inductances with an inductance modulated by the current from the electrochemical cell. Modern devices operate on the same principle as the original Oxford-Kew ozonesonde, except that the measurements of the electrochemical current are obtained digitally. Appreciable pre-launch preparation and calibration is needed (Figure 11.15).

11.4.3 Radioactivity and cosmic rays

Ionisation is generated near to the continental surface by natural terrestrial radioactivity, and, at higher levels, by cosmic rays. The temporal variation of ionisation with height has been measured using radiosondes equipped with Geiger tubes [109] by the Lebedev Institute in Moscow since 1957. A similar radiosonde approach can also be used to monitor environmental radioactivity [110] and ionisation variations in the upper atmosphere due to space weather changes [111].

11.4.4 Radiation

The standard sensors used to measure solar and terrestrial radiation, such as pyranometer and photodiode detectors are also amenable to use on radiosondes, but variation in the orientation of the instruments has to be considered or extra stabilisation included. Measurements using a combination of upward and downward facing pyranometers and pyrgeometers have used a radisoonde platform, stabilised by the use of two balloons, to derive clear sky profiles of long wave and short wave radiation [112].

Photodiode sensors provide a rapid response to changes in radiation. This presents the possibility of exploiting the motion of the balloon-radiosonde system for detection of cloud transitions, as the effects on solar radiation measurements differ inside and outside clouds. Figure 11.16 shows the marked change in solar radiation variability

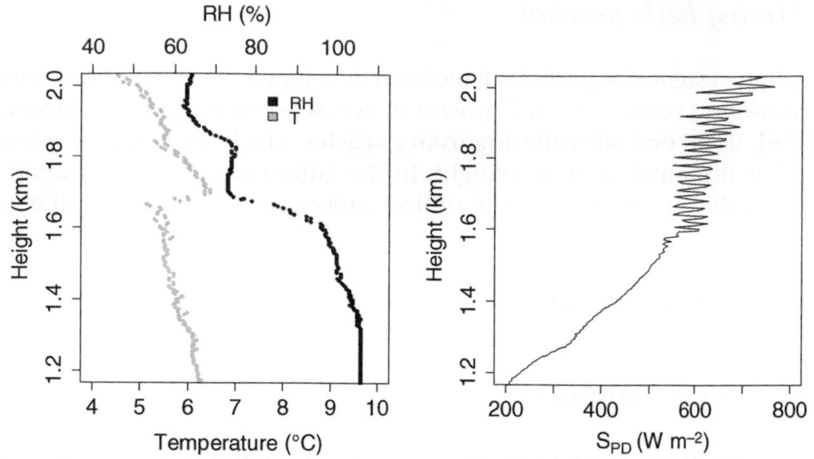

Figure 11.16 (a) Radiosonde measurements of RH (black points) and temperature (grey points), made at a cloud top at about 1.6 km. (b) Simultaneous solar radiation measurements (S_{PD}) made using a photodiode radiation sensor (redrawn from [31]).

occurring as a radiosonde passes through the clear air to cloud transition, when the solar radiation environment changes from strongly directional to diffuse and largely isotropic. The abrupt change at the cloud boundary in the measured radiation variability – from the motion of the radiosonde – also provides an indication of the slower time response associated with the capacitance RH sensor.

11.4.5 *Turbulence*

As well as responding to the mean wind, the motion of the radiosonde and balloon system also responds rapidly to smaller length scale fluctuations, such as those generated by turbulence, as well as the pendulum-like oscillations apparent from the radiation measurements above cloud in Figure 11.16b. These can be detected using a fast response motion sensors or miniature accelerometers. Magnetometers sensitive to the terrestrial magnetic field provide one method for sensing such fluctuations, as the variability measured in the magnetic field results from the platform's motion [113] and can be calibrated to variability in the vertical wind speed [114].

11.4.6 *Supercooled liquid water*

For clouds which contain supercooled droplets, the concentration of liquid water can be inferred from the amount of water freezing onto a sensing surface passing through the cloud. If the sensing surface used is in the form of a wire which is able to vibrate, the wire's natural oscillation frequency changes in relation to the mass of ice accreted, and therefore to the cloud liquid water concentration. This technique is very suitable for balloon-carried systems [115].

11.4.7 Atmospheric aerosol

Small sampling systems for particles have been developed for radiosondes, for example by encouraging condensational growth of aerosol so that optical transmission is reduced [116], or by optical scattering from particles which are pumped into a sampling chamber illuminated by laser light. In the latter case, size distribution information is available, as the number of optical pulses is proportional to the particle concentration, and the amplitude of the pulses related to the particle size. In combination with electrical measurements, aerosol particle counters have been used to monitor volcanic ash aloft [117].

11.5 Aircraft measurements

Commercial aircraft are equipped to make observations of pressure, height, air temperature, wind speed and direction, and routine aircraft measurements supplement the upper air measurements from the global radiosonde network. The aircraft obtaining these measurements form part of the World Meteorological Organisation's Aircraft Meteorological Data Relay (AMDAR) system. The aircraft measurement technologies now in use are briefly described, although they are frequently used in combination to allow corrections to be made, such as for air speed. A humidity sensor is not yet available, although methods based on absorption of light from a laser diode are under development. Some other properties can be inferred, such as turbulence, through comparison of the aircraft's orientation sensors with its navigation system. Beyond commercial aircraft, powered and unpowered gliders have provided very effective measurement platforms for certain conditions, such as investigating sea breeze fronts [118].

11.5.1 Air temperature

Temperature may be measured using a conventional sensor, usually electrical. However, any sensor deployed in the airstream will be subject to kinetic heating, as the air through which the thermometer is passing at high speed will be first locally accelerated to the speed of the thermometer, which is then decelerated almost to rest with local heating associated with the change in kinetic energy. The kinetic heating error is determined empirically for the thermometer *in situ* on the aircraft, or by wind tunnel calibration.

11.5.2 Wind

Commercial aircraft have precise navigation systems with which to provide ground velocity. An aircraft can also measure its velocity relative to the air, using a pitot tube (see Section 8.2.2) for relative airspeed, and by measuring the drift angle of the aircraft (the difference between true heading and apparent heading). The vector difference between these two velocities is the air velocity, typically measured with an accuracy of a few metres per second.

11.5.3 *Pressure*

This may be measured either directly from an aneroid barometer connected to a suitable static pressure head, or a pressure altimeter (an aneroid barometer calibrated in terms of height), again connected to a suitable static pressure head. Typical accuracy is ± 1 to 2 hPa.

11.5.4 *Altitude*

Commercial aircraft can be equipped with a radar altimeter, which measures the time of travel of a pulse of radio waves from the aircraft to ground and return. Typical accuracy is about ± 20 m.

11.6 Small robotic aircraft

New opportunities for controllable atmospheric measurements are now being provided by small autonomous aircraft, able to carry out predetermined flight plans using sophisticated navigation systems. These include fixed wing and rotor wing devices, in some cases having multiple rotors to offer good platform stability. At their smallest, the fixed wing aircraft have wingspans of under a metre and total mass around a kilogram (see Figure 11.17), and may therefore be considered effectively as a powered radiosonde. The propulsion is generally electric, using high-capacity lithium polymer batteries, which allows flight durations of tens of minutes reaching altitudes of several kilometres. These small platforms are variously known as *Unmanned Aerial Vehicles* (UAV), or *Remotely Piloted Aircraft* (RPA).

Figure 11.17 View of a small UAV carrying sensors for atmospheric measurements. An electrode to measure charge is mounted on the nose of the aircraft, and temperature sensors are carried within tubes visible alongside the fuselage above the wings. On the wings (not visible), are photodiode solar radiation sensors. (*Approximate platform specifications*: fuselage diameter 10 cm; wingspan 1 m; ceiling 4000 m and flight endurance 20 min.)

The computer systems and radio telemetry needed to fly such robotic aircraft are sophisticated, because of the need to trim the aircraft to keep it in level flight through responding to airspeed, attitude sensors and GPS in real time. The systems are generally designed so that carrying sensors for science measurements is straightforward, with, in contrast to radiosondes, substantial data rates possible and of course the final recovery of the sensor. As a consequence, new sensor technologies for these platforms are under active development [119].

12

Further Methods for Environmental Data Analysis

Environmental measurements tend to be motivated by specific science questions, which can also determine the analysis required. Conclusions about process, events and changes can be gleaned by using physical, conceptual or numerical models, statistical data descriptions or comparison with other measurement techniques. Combining measurements from multiple different instruments can also allow further quantities to be determined. Examples of possible data analysis are given here for a range of different types of measurements.

12.1 Physical models

The detailed understanding of processes associated with measured quantities ultimately improves theoretical descriptions and may also contribute to physical models of the processes concerned. Such physical models provide quantitative frameworks with which to compare and assess further measurements. In some cases, many factors have to be combined, leading to sophisticated computer models which couple different processes. The emphasis here is on physical and conceptual models which are numerically straightforward in permitting comparison with measurements, but illustrative of the approach which might also be taken with more elaborate numerical models.

12.1.1 Surface energy balance

At the earth's surface, the rates of arrival and removal of energy are in balance, at least on average. Measurement of the rate at which the net radiative energy is received from long wave and short wave radiation was described in Equation 9.1. The energy received is primarily dissipated by convection, evaporation or conduction (Figure 12.1), or through metabolic reactions if the surface forms part of a living

Meteorological Measurements and Instrumentation, First Edition. R. Giles Harrison.
© 2015 John Wiley & Sons, Ltd. Published 2015 by John Wiley & Sons, Ltd.
Companion website: www.wiley.com/go/harrison/meteorologicalinstruments

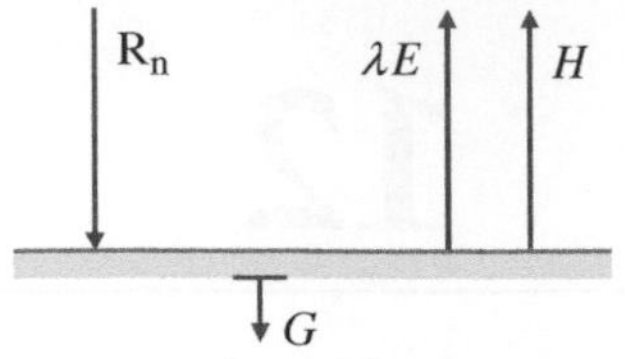

Figure 12.1 Terms in the surface energy balance. Radiative energy arriving at a rate R_n per unit area is balanced by turbulent transfer of heat (sensible heat flux H), and moisture (latent heat flux λE) and conduction (ground heat flux G).

object. Neglecting energy-exchange effects associated with precipitation, the equation[i] describing the energy balance at a flat surface is

$$R_n = H + \lambda E + G \{... \pm S\}, \quad (12.1)$$

where R_n is the net radiation arriving at an infinitely thin surface. (If the surface considered cannot be regarded as infinitely thin, which is the usual practical situation for a thin slab of soil or a vegetated surface, heat storage is also possible, and a further term S is needed to account for the energy passing to, or released from, storage.)

The principal terms on the right-hand side of Equation 12.1 describe the rate of transfer of energy away from the surface, per unit area. H represents the rate of removal of heat through turbulent transfer, known as the sensible heat flux density, and λE the heat transfer rate per unit area through moisture transport, where λ is the latent heat of vaporisation and E the mass transfer rate of water per unit area. The third term G is the ground heat flux density,[ii] which describes conduction beneath the surface.

Measurement of the individual terms in the surface energy balance depends on a combination of instrumentation techniques. Determining both R_n and G is straightforward in principle, utilising a net radiometer (Section 9.7) and ground heat flux plate (Section 5.6) respectively. The turbulent heat flux terms (H and λE) require more elaborate approaches. These fluxes can either be obtained assuming micrometeorological theory to allow calculation from standard measurements of vertical wind profiles, temperature profiles or temperature variability, or by direct measurement of the turbulent fluctuations causing the heat transfer. The first approach essentially uses micrometeorological methods, based on the theory of Monin and Obukhov which accounts for variations in atmospheric stability, although this is beyond the scope of the discussion here.[iii] As the second approach of direct measurement is primarily

[i] The sign convention used here is that the incoming radiative flux is positive, and the fluxes away from the surface are negative.

[ii] A shortening to 'heat flux' rather than heat flux density, is widely made.

[iii] See, for example, J.C. Kaimal and J.J. Finnigan. 1994. *Atmospheric Boundary Layer Flows: Their Structure and Measurement*, Oxford University Press.

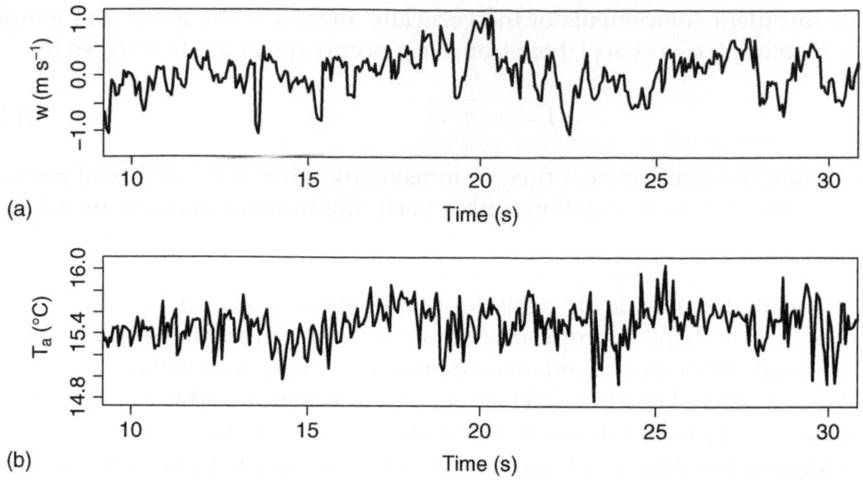

Figure 12.2 Simultaneous co-located micrometeorological time series of (a) vertical wind speed w and (b) air temperature T_a measured at Swifterbant, Netherlands (7 May 1994). (See also Figure 12.21.)

instrumental, using a combination of rapid-response instruments, it is now briefly considered further.

12.1.2 *Turbulent quantities and eddy covariance*

A key concept in surface layer data analysis is the description of turbulence through separation of the steady characteristics of a flow from its turbulent fluctuations. In the case of horizontal wind speed, such as for the data shown in Figure 12.2a, there is a well-defined mean wind speed, about which there are more rapid fluctuations. As a result, the variation of wind speed with time, such as the vertical wind speed $w(t)$, can be represented[iv] by combining the time-averaged wind speed $\overline{w}$ and time-dependent fluctuation $w'(t)$ as

$$w(t) = \overline{w} + w'(t)\,. \tag{12.2}$$

A similar decomposition can be made for temperature, humidity or indeed other quantities transported by the turbulence. In the case of temperature, the equivalent relationship for time averaged temperature $\overline{T}$ and time-dependent temperature fluctuation $T'(t)$ is

$$T(t) = \overline{T} + T'(t)\,. \tag{12.3}$$

In a time series of turbulent parameters, such as those shown in Figure 12.2, coincident fluctuations can be observed between the vertical wind speed and temperature.

[iv] This separation of the mean and fluctuating part of a quantity is also known as *Reynolds decomposition*.

When the turbulent components of the vertically aligned wind speed and temperature vary together (i.e. co-vary), heat transfer is occurring, at a rate H given by

$$H = \rho c_p \overline{w'T'}\,. \qquad (12.4)^{\text{v}}$$

Determining the sensible heat flux, or indeed any other vertical flux of moisture, gas or momentum if its association with a vertically moving mass of air is known, therefore becomes a problem of obtaining rapid fluctuations in the vertical wind speed fluctuations and the quantity concerned.[vi]

A sonic anemometer typically provides rapid measurements of wind speed at up to 100 Hz, which, with the appropriate alignment,[vii] can be used to determine the vertical wind speed. Associated rapid measurements of trace gas or water vapour can be obtained using infrared or ultraviolet absorption techniques, and fine wire resistance thermometers can be used to measure rapid temperature changes. If the sensor's temporal response is too slow for all the turbulent fluctuations to be resolved, or deteriorates with increasing frequency (Section 2.2.2), there will be an underestimation of the derived turbulent flux. Corrections can be applied if a sensor's response is known to be poor in the frequency ranges carrying the majority of the turbulent energy.[viii]

12.1.3 *Soil temperature model*

The diurnal variation of temperature at the soil surface responds to the diurnal cycle in solar radiation and therefore shows close similarities with the diurnal cycle in air temperature (see Figure 12.3a). With increasing depth into the soil, the surface temperature wave is attenuated in amplitude and retarded in phase (Figure 12.3b), as heat is absorbed or released along the path of thermal transfer. If the properties of the soil are constant with depth, the variation of temperature T with time t and depth z can be represented as

$$\frac{\partial T}{\partial t} = \kappa \frac{\partial^2 T}{\partial z^2}\,, \qquad (12.5)$$

where κ is the thermal diffusivity. κ depends on the soil properties, in particular the soil moisture. A simple solution to the heat transfer equation is

$$T(z,t) = T_m + A(z)\sin(\omega t - \phi(z))\,, \qquad (12.6)$$

[v] The over-bar here denotes averaging. It is evaluated by averaging the products of the fluctuations w' and T', that is, for N pairs of measurements, by calculating $\frac{1}{N}\left(w_1'T_1' + w_2'T_2' + \ldots w_N'T_N'\right)$.

[vi] A full description of this method is given in Eddy Covariance: A Practical Guide to Measurement and Data Analysis (Marc Aubinet, Timo Vesala, Dario Papale), pubiished by Springer in 2012.

[vii] Appendix B discusses how the three-dimensional sonic anemometer data from an instrument can be aligned along the flow directions.

[viii] The *eddy covariance* method described here represents the vertical flux at the point where coincident measurements are made; methods providing averaged properties along a path provide another approach. An approach to a path measurement is given by a *scintillometer*. This transmits and receives radiation, typically a modulated infrared signal, over distances of hundreds of metres. The sensible or latent heat flux can be found from the path-averaged variations in the refractive index of air observed, which result from temperature and humidity fluctuations.

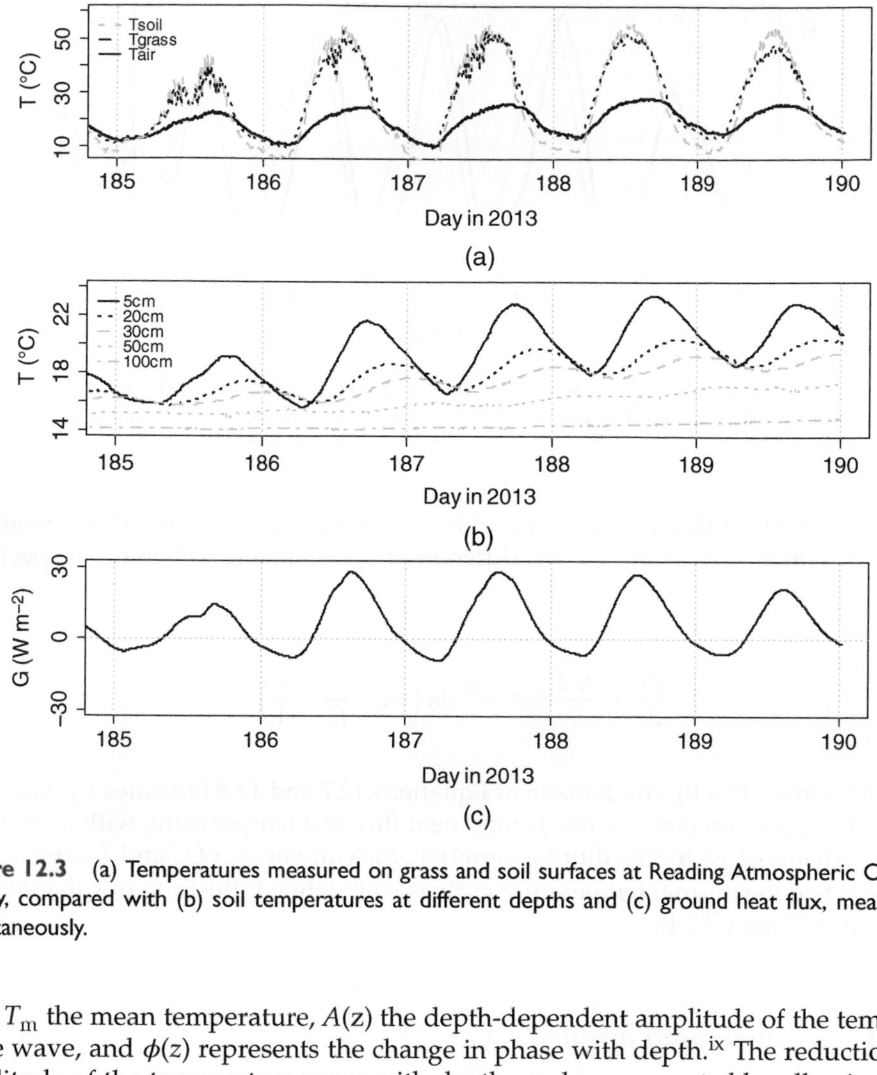

Figure 12.3 (a) Temperatures measured on grass and soil surfaces at Reading Atmospheric Observatory, compared with (b) soil temperatures at different depths and (c) ground heat flux, measured simultaneously.

with T_m the mean temperature, $A(z)$ the depth-dependent amplitude of the temperature wave, and $\phi(z)$ represents the change in phase with depth.[ix] The reduction in amplitude of the temperature wave with depth can be represented by allowing the surface amplitude to fall off with depth to $1/e$ of its surface value at a depth D, known as the *damping depth*. The phase shift can also be accounted for using this parameter, as

$$T(z,t) = T_m + A_0 e^{-z/D} \sin\left(\omega t - \frac{z}{D}\right), \tag{12.7}$$

where A_0 is the amplitude of the temperature wave at the surface.

In reality, the temperature variation is much more complicated than a single sinusoidal oscillation, with higher order harmonics also present. However, as Figure 12.3b shows, the change in phase and amplitude is similar to that expected from Equation 12.7.

[ix] The angular frequency ω for a single diurnal cycle is $2\pi/24$ radians hour^{-1}.

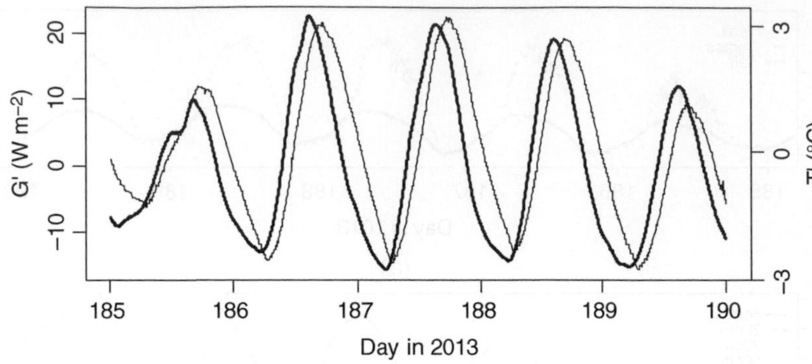

Figure 12.4 De-trended measurements of ground heat flux G' (thick line, left-hand axis) and soil temperature T' (thin line, right-hand axis), both at 5 cm depth, for summer days at Reading Observatory.

The ground heat flux G (Figure 12.3c) is proportional to the vertical temperature gradient, which can be found by differentiating Equation 12.7 with respect to depth as

$$\frac{dT}{dz} = \frac{\sqrt{2}}{D} A_0 e^{-z/D} \sin\left(\omega t - \frac{z}{D} + \frac{\pi}{4}\right). \tag{12.8}$$

At the surface ($z = 0$), comparison of Equations 12.7 and 12.8 indicates a phase difference between variations in the ground heat flux and temperature, with G leading T by $\frac{\pi}{4}$ radians, or 3 h for the diurnal variation. Measurements of G and T close to the surface when the diurnal temperature cycle is well defined illustrate this theoretical expectation (Figure 12.4).

12.1.4 *Vertical wind profile*

The variation of horizontal wind speed with height close to the surface shows a characteristic logarithmic form. This is a result of practical importance, as it allows wind speed measurements obtained by an anemometer at a non-standard height to be corrected (scaled) to the standard measured height of 10 m.

The origin of the logarithmic variation of wind speed with height is usually understood by considering the upward and downward displacement of air by turbulent eddies, Figure 12.5.

On this basis, if air is displaced downwards by turbulence from a height $z + l$ at which it had a horizontal speed $u(z + l)$ to a lower height z, where its speed is $u(z)$, then the turbulent change in velocity u' can be approximated as

$$u' = u(z + l) - u(z). \tag{12.9}$$

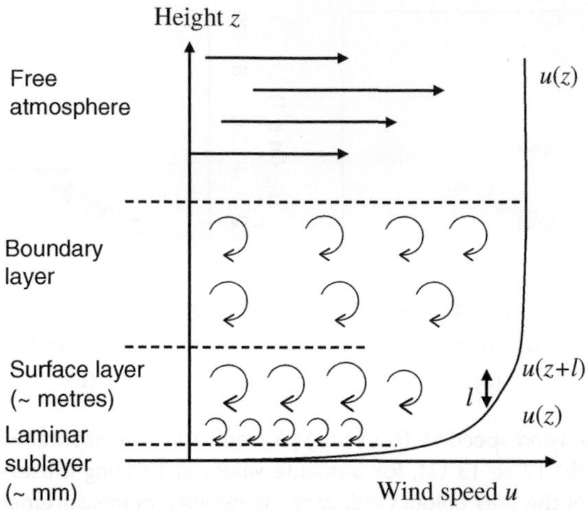

Figure 12.5 Variation of horizontal wind speed with height in neutral stability conditions, also depicting the increase of eddy scale above the surface. (The key atmospheric layers are marked.)

For l a length scale representative of the eddy, the speed fluctuation can also be expressed as

$$u' = l\frac{du}{dz}. \tag{12.10}$$

By assuming that the eddy sizes scale in linear proportion to their distance from the surface (i.e. $l = kz$),[x] and writing u_* as the typical size of u',

$$\frac{du}{dz} = \frac{u_*}{kz}, \tag{12.11}$$

which integrates to give

$$u(z) = \frac{u_*}{k} \ln\left[\frac{z-d}{z_0}\right], \tag{12.12}$$

with z_0 the constant of integration. z_0 is known as the *roughness length*[xi] of the surface. The consistent ratio between wind speed measurements at different heights implied by Equation 12.12 is clearly evident in wind speeds measured at 2 m and 10 m across the day, as shown in Figure 12.6.

---

[x] k is known as von Karman's constant ($k \sim 0.4$)

[xi] Roughness in this context is *aerodynamic*, that is it refers to an effect on the flow, rather than how the surface appears visually. Qualitatively, the roughness length of rugged heather moors is typically several orders of magnitude greater than the roughness length for a surface of flat ice. A zero-plane displacement height d is added to prevent a discontinuity at the surface. Physically, d is an effective height at which momentum is absorbed.

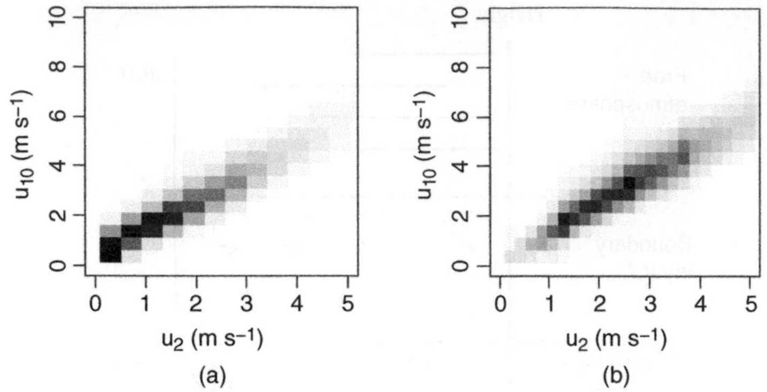

Figure 12.6 Horizontal wind speed at 10 m (u_{10}) plotted against horizontal wind speed at 2 m (u_2) for (a) 00 to 01 UT and (b) 12 to 13 UT, for 5-minute values at Reading Observatory between 2007 and 2012. (The intensity of the grey colours indicates the number of measurements.)

12.2 Solar radiation models

12.2.1 Langley's solar radiation method

An estimate for the top of atmosphere solar irradiance can be made on a clear, cloudless day, by using Langley's method. This determines the top of atmosphere solar irradiance by extrapolating the attenuating effect of the atmosphere to zero. Before satellites could measure irradiance directly outside the atmosphere, this was the only method available, for example using mountain-top sites to minimise aerosol effects.

The solar irradiance I at the base of the atmosphere is reduced from the total solar irradiance S_0 (Section 9.2.5) by attenuation in the atmosphere. It is given by Beer's law as

$$\frac{I}{S_0} = e^{-\delta}, \tag{12.13}$$

where δ is the optical depth along the path of the solar beam (see Figure 12.7).

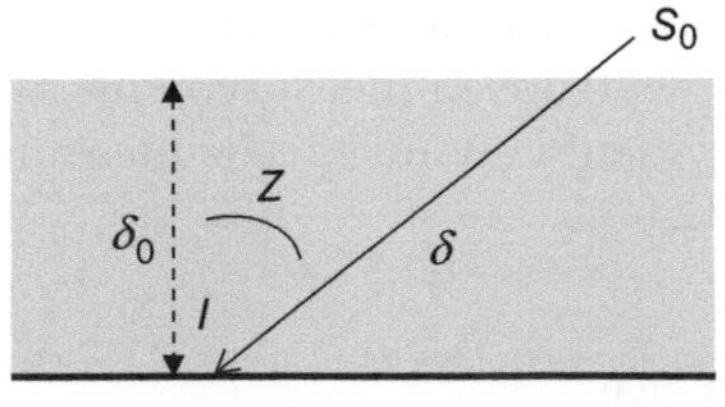

Figure 12.7 Solar geometry for the Langley method. The irradiance outside the atmosphere, S_0, is reduced to a value I at the surface by its passage along a slantwise path, which, if the vertical optical depth of the atmosphere is δ_0, has an effective optical depth δ. (Z is the solar zenith angle.)

Throughout the sun's transit during the course of a day, the optical depth along the path of the solar beam varies with the solar zenith angle Z, as $1/\cos(Z)$. This gives

$$\ln I = \ln S_0 - \delta_0 \sec(Z), \tag{12.14}$$

which indicates that a linear relationship can be obtained by plotting $\ln(I)$ against $\sec(Z)$. This method of plotting is useful as the gradient of the line then represents the vertical optical depth δ_0, and extrapolation to the intercept at $\sec(Z) = 0$ provides an estimate of S_0, the irradiance outside the atmosphere. It amounts to using the variation in solar angle to establish the variation in irradiance, and then extrapolating the optical depth to zero to estimate the irradiance without the effect of the atmosphere. An important requirement is that the atmospheric properties, such as the aerosol and water vapour content, must remain constant during the measurements, or the extrapolation will be inaccurate. Consequently, only a short period of measurements on a clear day, or measurements made at sufficient altitude for a reduced aerosol effect, are needed if the method is to work.

Figure 12.8 shows a Langley determination using measurements of the direct beam solar radiation S_b on a cloudless day, for which S_0 was found by extrapolation as (1246 ± 20) W m^{-2}. The error bounds (two standard errors) only represent the error in the linear fit, rather than an estimate of the limitations in the circumstances during which the measurements were made, which are evidently much larger as S_0 has been measured outside the atmosphere by satellite radiometer instruments as $\sim$1365 W m^{-2}.

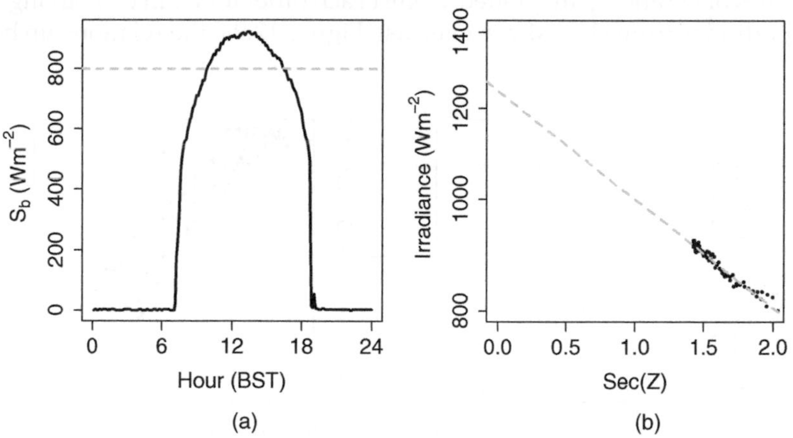

Figure 12.8 (a) Direct beam solar irradiance (S_b) measured at Reading on 7 September 2012 with a Kipp and Zonen CHP1 pyrheliometer (spectral range 0.2 μm to 4 μm), with the threshold solar radiation of 850 W m^{-2} marked with a dashed line. (b) Direct beam solar radiation for $S_b > 850$ W m^{-2} from the same day plotted as irradiance against sec(Z), where Z is the solar zenith angle. A least-squares fit line extrapolates the data to sec(Z) = 0, at which the irradiance $S_0 = (1246 \pm 20)$ W m^{-2}. The corresponding optical depth δ_0 derived from the slope is (0.0947 ± 0.004). (Error range from the fitted line represents two standard errors in each case.)

12.2.2 *Surface solar radiation: Holland's model*

Variations in measured radiation can be analysed in a different way in cloudy conditions. Measurements at the surface of the global solar irradiance S_g and diffuse solar irradiance S_d differ from the calculated top of atmosphere solar irradiance S_{TOA} because of absorption and scattering of the radiation during its passage through the atmosphere. In principle, therefore, the combined surface radiation measurements contain information about the radiative properties of the atmosphere above.

To study this further, two parameters can be derived from S_d and S_g. These are the *diffuse fraction*, DF, given by

$$DF = \frac{S_d}{S_g}, \quad (12.15)$$

and the *clearness index* K_t, given by

$$K_t = 1 - \frac{S_g}{S_{TOA}}. \quad (12.16)$$

K_t provides a measure of the attenuation which has occurred, and DF a measure of the associated scattering. As the amount of scattered radiation increases, DF varies from 0.2 in clear sky to a limiting value of 1.0 in overcast conditions (when all the radiation is diffuse), and K_t decreases from about 0.8 in clear conditions to close to 0 in heavily overcast conditions. After DF has reached its limiting value in overcast conditions, K_t continues to vary with cloud thickness.

If the values of DF and K_t are plotted against each other for a full day having a range of cloud conditions from clear sky to overcast, Figure 12.9a, the relationship between

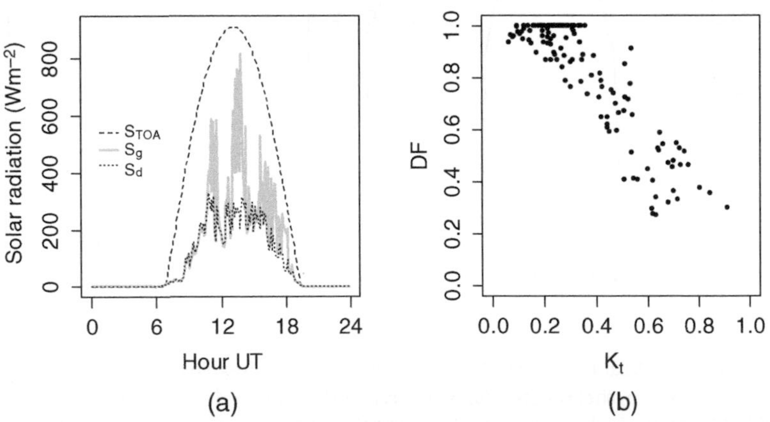

Figure 12.9 (a) Solar radiation measured at Reading on 14 September 2012 as 5-minute average values of global solar irradiance (S_g, grey line) and diffuse solar irradiance (S_d, dotted black line). The calculated top of atmosphere irradiance on a horizontal surface (S_{TOA}, dashed black line) is also shown. (b) Derived radiation quantities of Diffuse fraction ($DF = S_d/S_g$) and Clearness index ($K_t = 1 - (S_g/S_{TOA})$), plotted against each other.

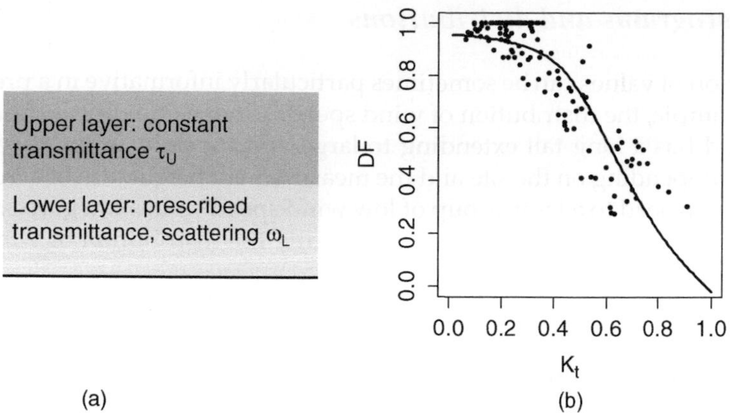

Figure 12.10 (a) Assumptions of the two-layer model. (b) Diffuse fraction DF plotted against Clearness index K_t, with (solid line) a statistical fit to Equation 12.17. (Derived parameters are $\tau_u = 0.981$ and $\omega_L = 0.979$.)

DF and K_t takes on a characteristic form, Figure 12.9b. This shows the 'saturation' of DF in the overcast conditions while K_t still varies, and the variation of DF with K_t in increasingly broken cloud conditions.

A theoretical approach which can describe the variation in Figure 12.9b was suggested by Hollands [120]. This model neglects surface reflection, and represents the atmosphere by two homogenous layers which are radiatively non-selective in wavelength. The upper layer is represented by a constant transmittance τ_u, with no scattering permitted, and a lower layer characterised by a constant scattering (specified by an albedo ω_L) and a prescribed variation in transmittance (Figure 12.10a). With these assumptions, an analytical result can be obtained to relate the diffuse fraction (DF) with the clearness index (K_t) as

$$\mathrm{DF} = \frac{\left[1 - b - \sqrt{(1-b)^2 - 4ab^2 K_t (1 - aK_t)}\right]}{2abK_t}, \tag{12.17}$$

where $\tau_u = 1/a$ and $\omega_L = 2b$. Figure 12.10b shows the data of Figure 12.9a fitted statistically to Equation 12.17, with a and b allowed to vary. A fair representation of the data can be achieved using this functional form.

12.3 Statistical models

In some cases, no simple physical model is available, and a solely statistical approach is needed instead. This may require the fitting of a statistical distribution to data from which characterising parameters can be found, or through the use of statistical tests to establish relationships between one quantity and another.

12.3.1 Histograms and distributions

The distribution of values can be sometimes particularly informative in a preliminary study. For example, the distribution of wind speeds is not symmetrical about its central value, and has a long tail extending to large wind speeds. In the case of mean wind speeds, depending on the site and the measurement height, use of a mechanical anemometer may lead to a large group of low wind speed values categorised as calm. Hence the choice of instrument can further modify the distribution of wind speeds obtained.

12.3.2 Statistical tests

The use of histograms to represent data can help to demonstrate whether the values from two different instruments are in fact different. This is ultimately formalised in statistical tests [121], which establish whether a particular hypothesis can be supported by the data, or whether it is more likely that the situation described by the data has arisen by chance.[xii]

Figure 12.11 shows examples of daily pressure measurements P1 and P2 made simultaneously at the same site using two different instruments. (The P1 values were previously presented as a time series in Figure 4.10.) If the two sets of values are plotted against one another (Figure 12.11a), some scatter around the straight line relationship expected is evident. Whether this scatter constitutes a significant difference between the datasets in a statistical sense can be investigated by comparing the distributions of P1 and P2. Figure 12.11b shows the distributions of the two quantities, each summarised using a boxplot outline. Figure 12.11c and d show the data from P1 and P2 as histograms plotted above each other. Whilst there are small differences apparent, both the medians and means are identical and the quartiles are very similar.

The comparison of two distributions of values in general is made using statistical tests, which are used to investigate if a particular property differs between the distributions. Many such tests are available, and a few are summarised in Table 12.1; they can be used separately or successively. For the two distributions of Figure 12.11, these tests show that it is not possible to reject the hypothesis that the mean values, median value or distributions are the same.

An example of the combined use of statistical tests is given in Figure 12.12, in which the daily averages of pressure (P), horizontal (u_{10}) and vertical (w_{10}) wind speeds obtained over a long period are compared. Figure 12.12a shows that the initial distribution of all available daily values of w_{10} is asymmetric, with the lower and upper quartiles of w_{10} at −1.6 cm s^{-1} and 4 cm s^{-1}. The origin of this asymmetry is more apparent from Figure 12.12b, which indicates a relationship between u_{10} and w_{10}, both positive and negative w_{10} values associated with increasing u_{10}. (This may arise

[xii] A statistical test is used to reject a null hypothesis, leaving the possibility that an alternative hypothesis is true. The confidence with which the null hypothesis can be rejected is measured using a p-value provided by the test. If the p-value is 0.05 or smaller, there is only a 5% (or smaller) probability that the outcome of the test was obtained by chance and hence the result is generally regarded as *significant* (although this does not mean that there is a 95%, or better, probability that the null hypothesis is false). Small p-values are associated with an increasing level of significance at which the null hypothesis can be rejected.

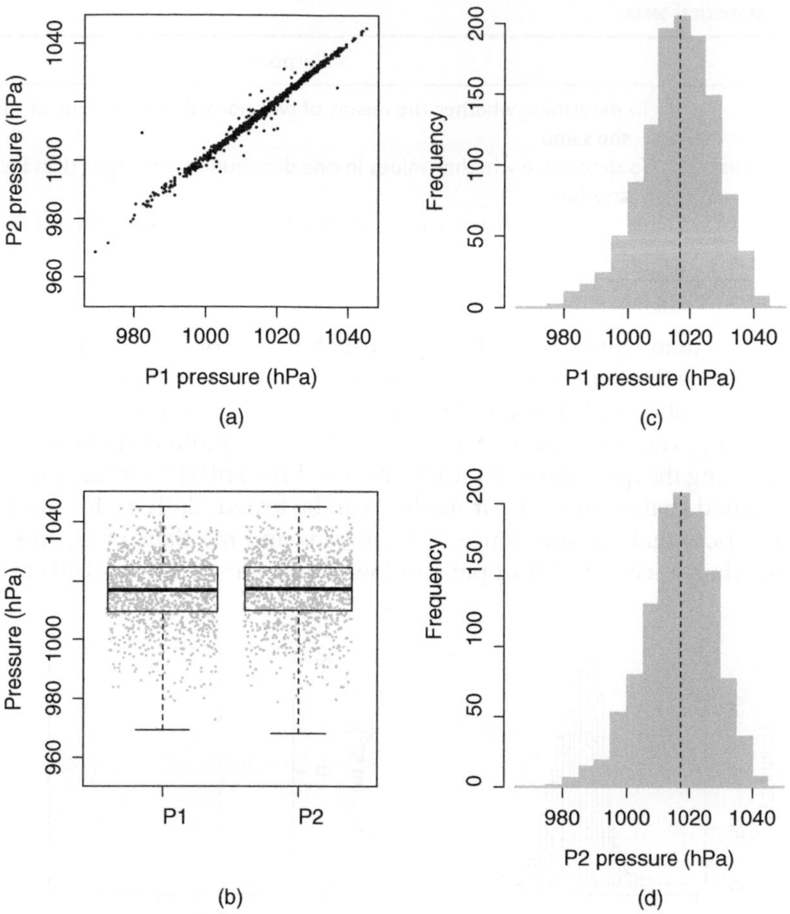

Figure 12.11 Comparison of two sets of simultaneous daily station pressure measurements, using two different instruments P1 and P2 situated at the same site. (The P1 values are from a mercury barometer, and the P2 values from an adjacent vibrating drum barometer). (a) shows the two sets of values plotted against each other. (b) shows the two sets of data plotted as distributions, with random horizontal scatter added for clarity. A 'boxplot' outline has also been drawn around the points to summarise their distribution: this shows the median value as a thick line, the interquartile range as the outline of each box, and uses whiskers to mark the extreme values. (c) and (d) provide the associated histograms of the P1 and P2 data with the median value marked (dashed line).

from the contributions of the horizontal flow to the vertical flow, such as from incompletely aligned horizontal flows.)

If the data values are restricted to days on which the horizontal wind speed is small ($u_{10} < 2$ m s^{-1}), Figure 12.12c shows that the distribution of w_{10} becomes narrower but more symmetric, with lower and upper quartiles -1.55 cm s^{-1} and 1.52 cm s^{-1}. Any remaining sensitivity of u_{10} to w_{10} can be evaluated by using statistical tests. For example, if the w_{10} data are divided into days on which u_{10} was above and below its median value, a t-test (or the Mann-Whitney test) can be used to evaluate if the mean (or median) values of w_{10} are different. In this case, the statistical tests show that both

Table 12.1 Statistical tests

Statistical test	Purpose
t test	To determine whether the means of two normal distributions of values are the same
Mann–Whitney test	To determine whether values in one distribution are larger than those in another
Kolmogorov–Smirnov test	To determine whether two sets of values are drawn from the same distribution

the mean and median values of the two w_{10} distributions for large and small u_{10} are not different, hence the dependence of w_{10} on u_{10} has been removed.

With no appreciable contribution of u_{10} to w_{10}, any remaining sensitivity between pressure and w_{10} can now be considered and w_{10} is plotted against P in Figure 12.12(d). Using the quartiles of P (at 1005 hPa and 1018 hPa) to divide the values of $w_{10,}$ an associated change in mean or median can be tested. Both a *t*-test and Mann–Whitney test show that the difference in the means and medians of w_{10} are significant, with the daily mean w_{10} for upper and lower quartiles of P are -0.40 cm s^{-1} and

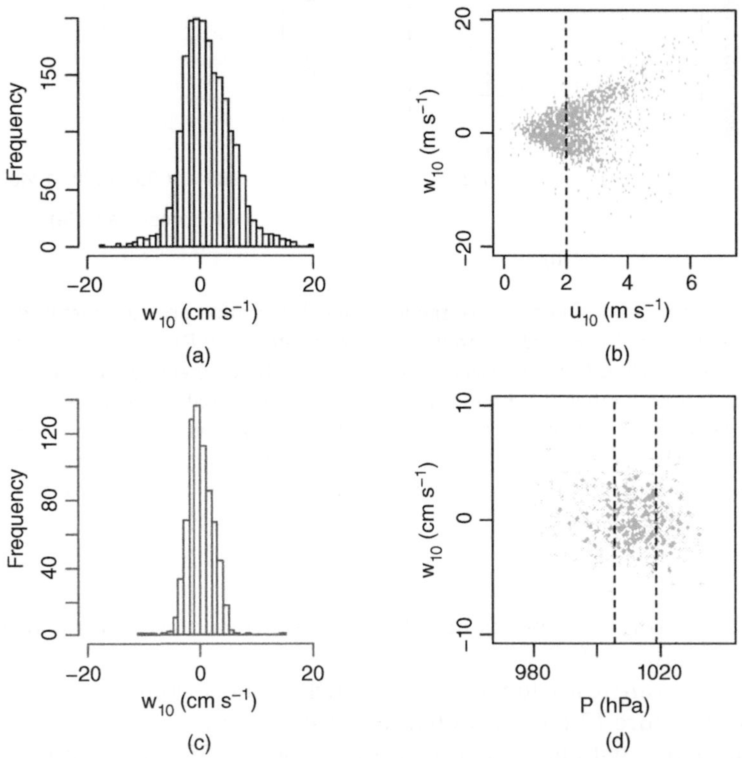

Figure 12.12 Daily averages of vertical and horizontal wind speeds measured with a sonic anemometer at 10 m (denoted as u_{10} and w_{10} respectively, with w_{10} positive downwards), for 6 years. (a) shows the w_{10} distribution. (b) shows u_{10} and w_{10} plotted against each other, with $u_{10} = 2$ m s^{-1} marked. (c) shows the distribution of w_{10} values sub-selected for $u_{10} < 2$ m s^{-1}. (d) Sub-selected w_{10} values from (c), plotted against daily mean pressure, P, with the upper and lower quartiles of P marked.

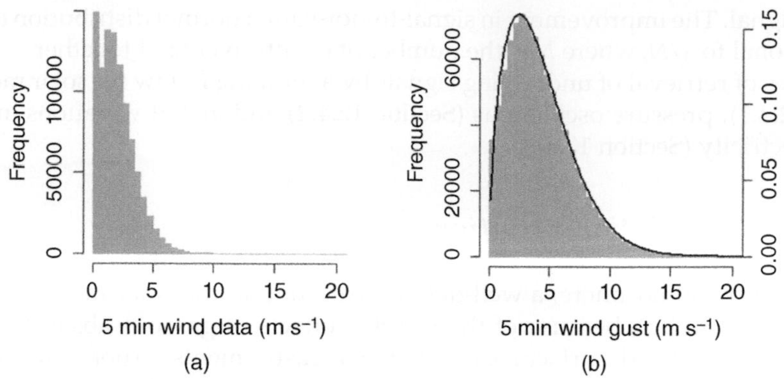

Figure 12.13 Distribution of wind speeds measured at Reading between 1997 and 2006, at 2 m above the surface using a cup anemometer, (a) Mean 5-minute wind speed, and (b) Peak 5-minute wind speed. A Weibull distribution has been fitted to the data in (b) (black line, with associated probability density on the right-hand axis), for which the shape factor k is 1.57 and the scale factor λ is 4.75 m s^{-1}.

0.38 cm s^{-1} respectively. Beyond providing data summaries, distributions of specified kinds can also be fitted to data, as now illustrated for the case of wind gusts.

12.3.3 *Wind gusts*

Wind gusts are of engineering importance because they can cause structural damage. Extreme values of wind speeds occur only rarely in a set of measurements by definition and are therefore not well sampled. For this reason gust information may have to be estimated, such as by fitting a statistical distribution to wind data and the probability of wind speeds occurring above a given threshold calculated. The design of buildings or structures may then be considered in terms of the statistical properties of the wind at the site, which can indicate the probability of extreme values. A Weibull probability distribution is widely used to represent wind gusts. For positive values, its probability density function $f(x)$ is defined by

$$f(x) = \frac{k}{\lambda}\left(\frac{x}{\lambda}\right)^{k-1} \mathrm{e}^{-(x/\lambda)^k}, \tag{12.18}$$

where k is a shape parameter and λ a scale parameter. (An example of the use of this distribution is provided in Figure 12.13.)

12.4 Ensemble averaging

For measurements which show considerable variability on short time scales, an underlying signal can sometimes be extracted through averaging multiple events of a similar kind together.[xiii] This *ensemble averaging* acts to improve the signal-to-noise ratio, as it reduces noise which is uncorrelated between events, whilst reinforcing any

[xiii]Averaging around defined triggering events is known as *compositing*, or *superposed epoch analysis*. If there are enough similar events, this approach can isolate the characteristic effect of the triggering event.

common signal. The improvement in signal-to-noise for a normal distribution of noise is proportional to $\sqrt{N}$, where N is the number of events averaged together.

Examples of retrieval of underlying signals by averaging follow for solar radiation (Section 12.4.1), pressure oscillations (Section 12.4.2) and global variations in atmospheric electricity (Section 12.4.3).

12.4.1 *Solar radiation variation*

At the top of the atmosphere, a well-defined variation in the solar radiation occurs (see Section 9.2.5), but, because of the variable effects of gaseous absorption, and, more importantly, cloud, surface solar radiation measurements do not show the same well-defined variation except on unusually clear days. Figure 12.14 shows the theoretical variation expected on average across all days, and the effect of averaging randomly selected days' values together, for an increasing number of days. Only when very many days are averaged together (5000 days in Figure 12.14d) is the smooth curve expected retrieved. Its mean value differs from the theoretical curve because of absorption and cloud effects which are also averaged, but the hour-to-hour

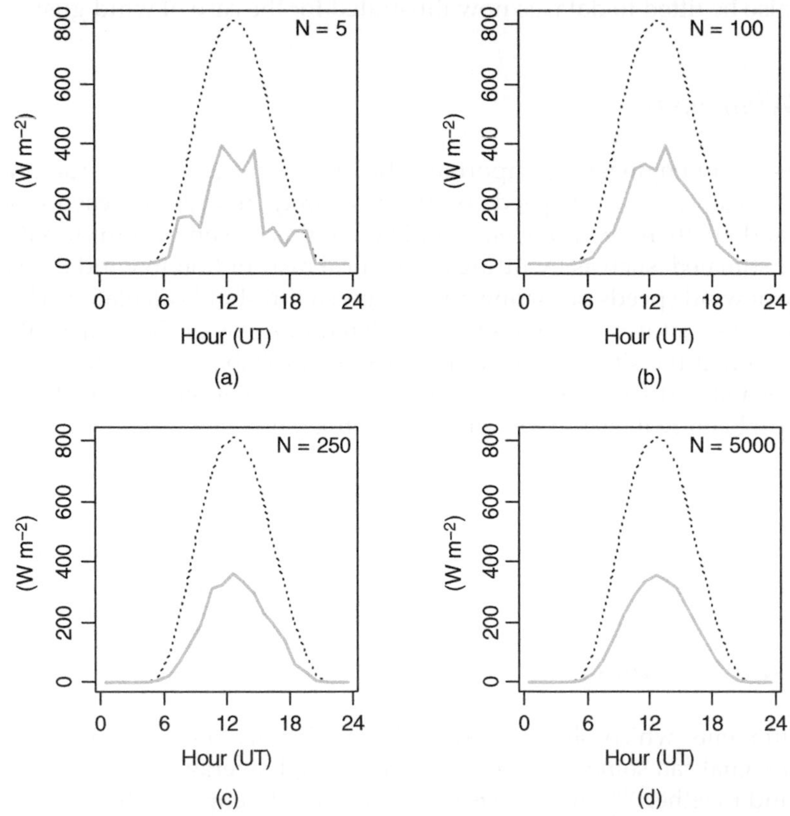

Figure 12.14 Effect of averaging solar radiation values (S_g) samples obtained at the same hour of day (grey lines), for (a) 5, (b) 100, (c) 250 and (d) 5000 randomly selected hourly values from 3 years of measurements at Reading. The dashed line shows the calculated top of atmosphere solar irradiance, averaged at each hour across the year.

variability is removed by the averaging. In this way, the essentially random variability generated by cloud between the top and base of the atmosphere is averaged away, allowing the driving diurnal variation to emerge.

12.4.2 *Pressure tides*

Another example of the reduction of random variations by averaging to reveal an underlying variation is the analysis of surface pressure values. Global data shows small (~1 hPa) diurnal and semi-diurnal oscillations in pressure, generally known as tides, which result from a complex combination of solar heating and wave generation. In mid-latitudes, the diurnal oscillation has its maximum around 1000–1200 Local Time (LT), and the semi-diurnal maxima occur around 1000 and 2200 LT [122]. At times when the surface pressure is very steady (Figure 12.15a), this small variation can be seen directly in barometer measurements if there is sufficient resolution in the instrument. An alternative method is to average hourly pressure values over about a year [123], which, after the mean value is removed, shows the amplitude and phase of the pressure oscillation (Figure 12.15b).

If a longer series of pressure values is available, the random variability will be more effectively reduced (Figure 12.16a), although the variation is similar in phase and amplitude to that obtained in Figure 12.15b. With many more values available, averages can also be calculated by season, allowing the slight variation in phase of the oscillation to be retrieved, most clearly present in the secondary minimum around 16 UT (Figure 12.16b).

12.4.3 *Carnegie curve*

A further example of an underlying diurnal variation usually obscured by local fluctuations occurs in atmospheric electricity, in which a variation linked to universal time (rather than local time at a site) results from the combined effect of the daily variation of global thunderstorms. This is known as the *Carnegie curve* [124], after the

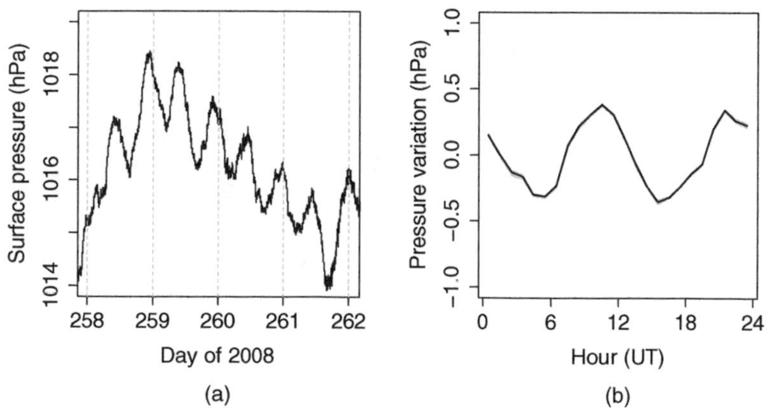

Figure 12.15 (a) Variation in surface pressure measured at Reading during September 2008 (1 minute data), with day boundaries marked. (b) Hourly fluctuations in pressure, found from averaging the data in (a).

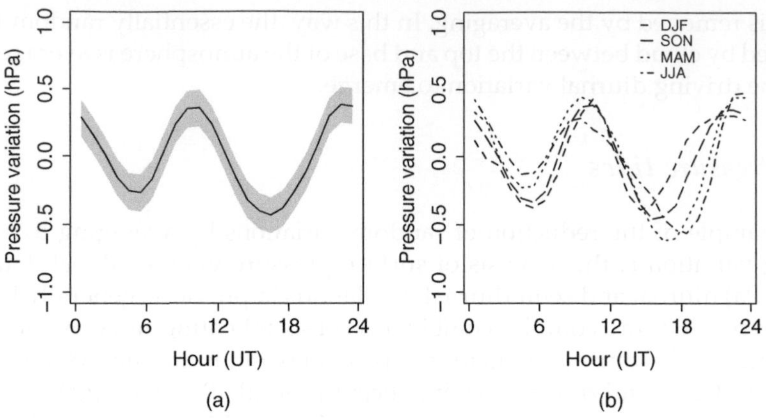

Figure 12.16 Hourly averages of surface pressure (a) across all 5-minute samples 2007–2012 (shaded band shows the range of 1.96 standard errors on the mean value) and (b) the mean values found by season (winter: December–January–February, spring: March–April–May, summer: June–July–August, autumn: September–October–November).

ship on which the original defining measurements were made. This underlying variation is, unlike the pressure and solar radiation variation, global in origin and extent. (It is also sensitive to internal variability of the climate system, such as through the El Niño oscillation [125].) Without averaging, however, it is almost always obscured, and hardly ever apparent in a single day's measurements.

Figure 12.17a shows the daily variation in Potential Gradient, averaged at each hour from surface measurements obtained at Lerwick Observatory over a number of years [126]. The phasing of this diurnal oscillation is very similar to the variation in global thunderstorm area shown in Figure 12.17b, as expected physically through the global flow of electric current.

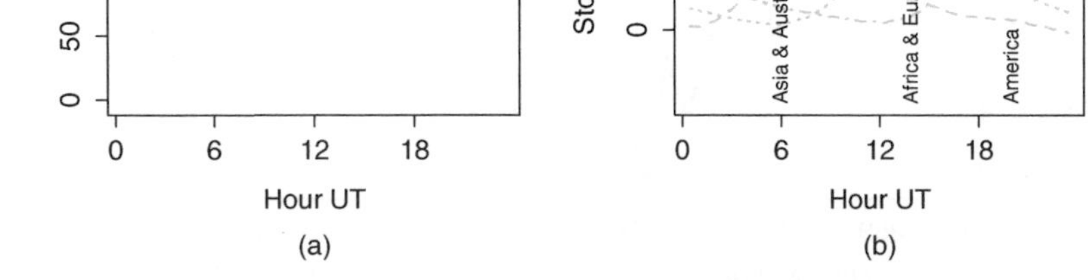

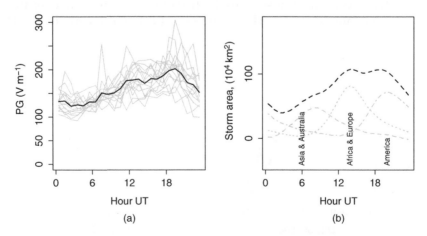

Figure 12.17 (a) Hourly averaged Potential Gradient (PG) at Lerwick in December, for each of the years 1968 to 1984 (grey lines), and averaged across all the years (thick black line). (b) Hourly averaged variation in global thunderstorm area, with the principal contributions from three different geographical regions identified.

12.5 Spectral methods

12.5.1 *Power spectra*

If different characteristic underlying periodicities are expected to be present, one method to identify them is through calculation of a *power spectrum*, which determines the amplitude of the contributions from the different periodicities present. The power spectrum is usually calculated by the numerical procedure of a Fourier Transform if the data values are regularly spaced.[xiv] A particularly efficient method of calculating a Fourier Transform is known as the *Fast Fourier Transform*, or FFT. Using the FFT, the highest frequency which can be resolved is the Nyquist frequency (Section 4.1.5).

A further aspect immediately evident from the power spectrum is the change in the spectral power density from low frequency (long period) to high frequency (short period). A negative slope on a spectral plot against increasing frequency is characteristic of many geophysical quantities, and is generally known as a *red* spectrum, by analogy with a visible spectrum in which the longer (red) wavelengths dominate. Using the same analogy, when the spectral power is equal at all frequencies, the spectrum is said to be *white*.

Some geophysical data show *persistence*, in that variations are influenced to some extent to earlier variations. Persistence in a geophysical data series may not be at all evident from the original time series, but can be readily determined from the power spectrum, as illustrated using synthetic data in Figure 12.18. A *spectral slope* indicates that some frequencies are more strongly present than others in a time series, and in the case of a red spectrum, that some structure exists, in the form of longer period (lower frequency) processes which dominate. In contrast, a white spectrum arises when all the data values are independent of each other.

The extent of persistence in a time series can also be evaluated using *autocorrelation*, through calculating the correlation between groups of points in a time series which are increasingly separated in time. For random events showing no persistence, there will be negligible correlation between subsequent values; if there is persistence, the autocorrelation will decrease as the time interval used for the correlation exceeds the timescale within which one value can influence another.[xv]

Figure 12.19 shows a power spectrum calculated from the pressure data presented in Figure 12.15a, in which it is already known – both by eye and by averaging – that a semi-diurnal oscillation is present. Accordingly, the power spectrum shows peaks at both semi-diurnal and diurnal periods.

Because of the temporal structure commonly present in geophysical time series, a white power spectrum, or its approximation, is rare. In order for a spectrum to be white, the information contained must become randomised to remove the temporal structure. This is sometimes seen in the time series of diffuse solar radiation on cloud-free days, as a result of the multiple scattering processes occurring to the solar

---

[xiv] For observations irregularly sampled in time or in which some are missing, the *Lomb periodogram* method can be used instead. Example programs for this and the FFT are given in *Numerical Recipes: The Art of Scientific Computing* (Cambridge University Press, 2007).

[xv] There are efficient computing approaches to calculating the autocorrelation using the FFT, which are particularly suitable for large datasets. These are based on the Wiener–Khinchin theorem, which shows that the Fourier transform of the autocorrelation is equivalent to the power spectral density.

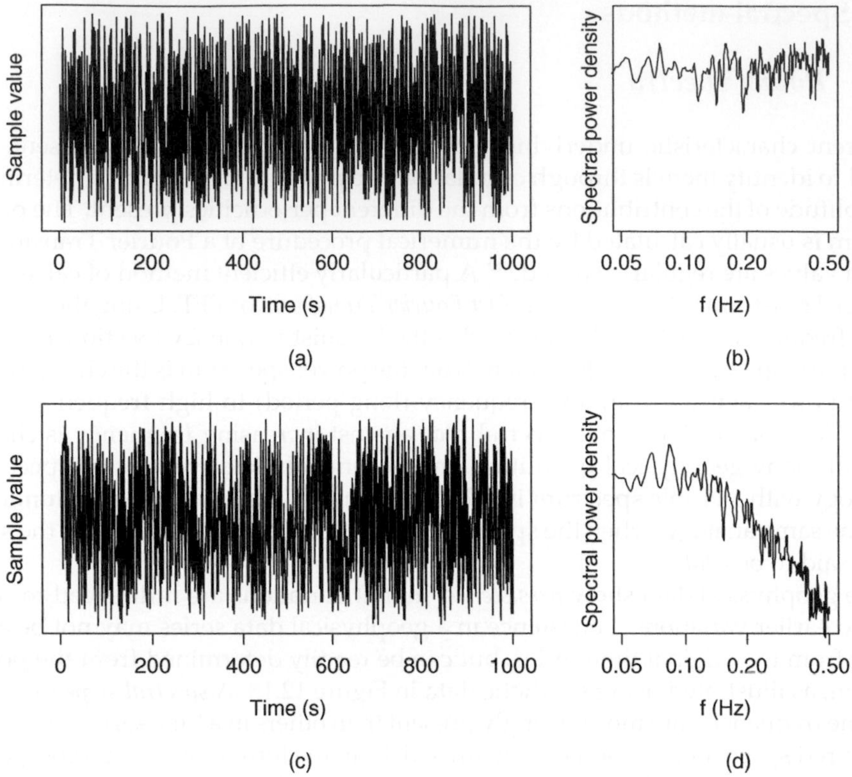

Figure 12.18 (a) Time series of random values sampled every second with (b) the derived power spectrum. (c) Time series of random values in which each value contains a 10% contribution from its previous value, with (d) the associated derived power spectrum.

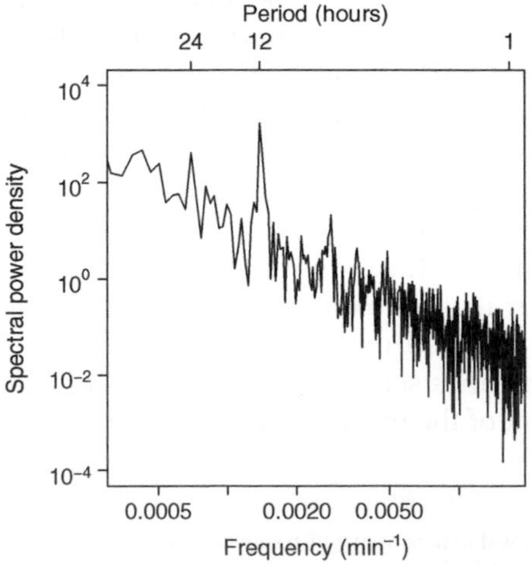

Figure 12.19 Power spectrum calculated from the pressure data shown in Figure 12.15a, using a Fourier Transform method. The relative spectral power present at different periods is shown, with the semi-diurnal (12 h) and diurnal (24 h) periods evident.

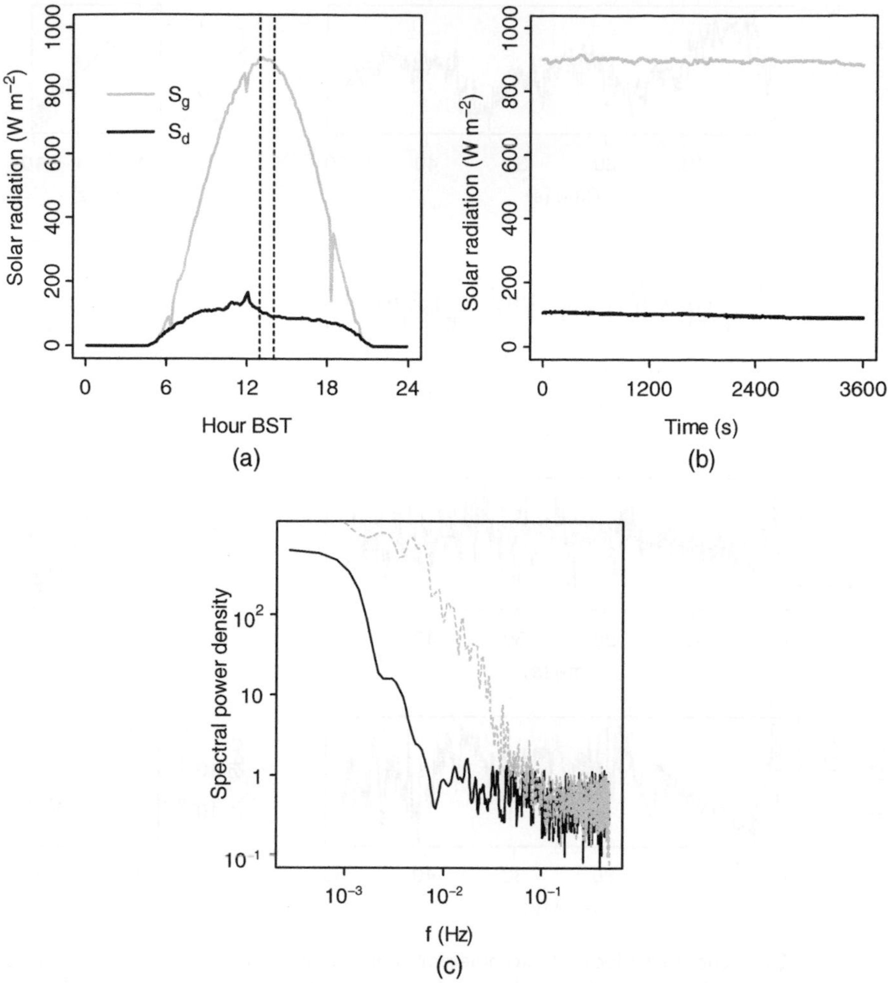

Figure 12.20 (a) Solar radiation Measurements at Reading of diffuse solar irradiance (S_d, black line) and global solar irradiance (S_g, grey line) at 1-second sampling on a cloudless day (9 June 2006), and (b) near local noon, 13 BST to 14 BST. (c) Power spectra for S_g and S_d, from the data in (b). In each case, the black lines are from (Identical Kipp and Zonen CM5 pyranometers were used, with exponential time constants ~10 s.)

radiation passing through the atmosphere. In contrast, the surface global solar irradiance received has a much smaller proportion arising from scattering, and more structure remains. Figure 12.20 shows the power spectra for S_g and S_d (obtained using a shade band) around noon on a clear summer day. Despite identical instrumentation, the power spectra differ and the power spectrum for S_d is flatter in the range 0.01 to 0.1 Hz (i.e. more readily considered as 'white') than that for S_g.

12.5.2 *Micrometeorological power spectra*

For atmospheric quantities such as specific humidity, wind speed and temperature which can be sampled rapidly near to the surface, similar spectral slopes are seen

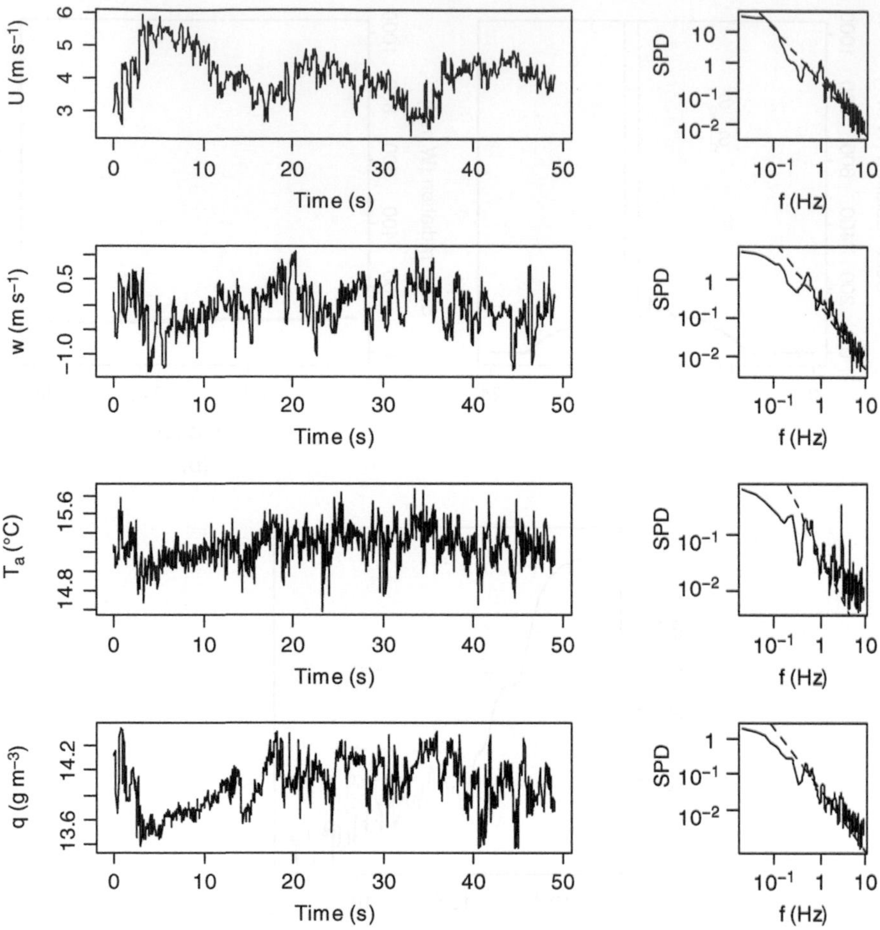

Figure 12.21 Simultaneous co-located micrometeorological time series of (left-hand panels, top to bottom): mean horizontal wind speed *U*, vertical wind speed *w*, air temperature T_a and specific humidity *q*. The Spectral Power Density (SPD) has been computed for each of the time series (right-hand panels). (Measurements obtained at 20 Hz at Swifterbant, Netherlands, on 7 May 1994, using a sonic anemometer, fine wire platinum resistance thermometer and infrared hygrometer.) On each power spectrum, a spectral slope of −5/3 is shown with a dashed line.

at high frequencies. This is because all the quantities are influenced by surface layer turbulence. Although the nature of the turbulence depends on the local atmospheric stratification, the turbulent energy is transferred to high frequencies by break up of successively smaller eddies. The slope of the spectral power density for motions in this range[xvi] is predicted by Kolmogorov's theory [127] to be −5/3. As temperature and humidity are carried by the flow, they are also influenced by these spectral variations. Figure 12.21 shows short simultaneous measurements of wind speeds, air temperature and humidity, with the respective power spectra derived. Spectra for

[xvi] This spectral region is known as the inertial sub-range, through which energy passes from the large scales to small scales before its final viscous dissipation.

horizontal and vertical wind speeds show good agreement with the −5/3 Kolmogorov slope. This spectral slope is also approached by temperature and humidity spectra as the frequency increases.

12.6 Conclusion

Measurements and their analysis provide the basic quantitative environmental information with which to reinforce ideas, validate physical theories or determine when a critical threshold is reached, following careful characterisation, calibration, signal processing, sampling and recording of the sensors concerned.

Beyond the value of single point measurements, opportunities now exist to network measurements in new ways, through the use of wireless sensors, or the additional interconnectivity provided by the internet or satellite communication systems. The extra spatial information generated is, in principle, no different to that of early meteorological observing systems, but of course the amount of data which can now be obtained and the rate at which it can be disseminated is vast by comparison.

The response of an individual sensor to an environmental parameter will always be essential to characterising the physical world, and, within the larger and more elaborate measurement networks which can confidently be anticipated, the role of measurements and instrumentation can only become ever more important.

A

Writing a Brief Instrumentation Paper

A thorough description of the characterisation of an instrument is an essential prerequisite for a scientific study in which it is used. Such a primary source of information is important, as, if the instrument itself is well described and summarised, the scientific results obtained from its use ought to be subject to fewer doubts, ultimately giving more confidence in the overall findings. More pragmatically, if an instrument has been designed and made specifically for a particular investigation, without such a publication there may otherwise be no record of its existence.

A.1 Scope of an instrument paper

An instrument description can usefully be separated from the study in which it is used, either by putting the instrumentation aspects in an appendix to the main paper, or, if it contains original material, by publishing a self-contained short paper (often called a Note) describing the instrument and the tests and calibration checks made. Novelty is needed for the description to be publishable in this way, in an instrumentation journal such as *Reviews of Scientific Instruments* or *Measurement Science and Technology*.

A.2 Structure of an instrument paper

An instrument Note would generally consist of the following elements: *Title, Abstract, Motivation, Instrument description, Calibration* or comparison with another instrument, *Summary*. A few relevant references to earlier work are also essential for placing the novelty of the new device or approach in context. The different elements of an instrument note are now briefly summarised.

A.2.1 Paper title

A well-chosen title serves to define the content of an instrumentation paper and ideally should give the scope and relevance of its content. For example, beginning the

Meteorological Measurements and Instrumentation, First Edition. R. Giles Harrison.
© 2015 John Wiley & Sons, Ltd. Published 2015 by John Wiley & Sons, Ltd.
Companion website: www.wiley.com/go/harrison/meteorologicalinstruments

title with 'A balloon-borne ... ' or ending with '... for atmospheric temperature measurements', gives immediate context and the likely application.

A.2.2 Abstract

The Abstract is a short summary of the content of a paper. After a paper's publication, its Abstract is likely to be discovered through using search tools, and depending on the journal's publication policy, it will probably be available free. Combined with the title, the Abstract therefore forms an important aspect of the marketing of the final published work to a broad audience.

A typical Abstract is usually 100 to 200 words in length. Generally the Abstract can therefore only consist of about half a dozen sentences, the first couple giving the motivation and context, the second couple summarizing the evidence supporting the function of the device, quantitatively if possible, and a final couple describing the likely application and any limitations. Conciseness in the Abstract is important, through the use of short sentences and punctuation to separate different ideas (although using too many adjectives or hyphens can make it difficult to read). Numerical quantities need to be presented carefully, with units and uncertainties, for example 'The sensitivity of the sensor was (3 ± 1) V K^{-1}... ' An Abstract almost never contains references, except for some journals which specifically require an introductory paragraph rather than an Abstract, or (and this is rare), when the work of one particular previous paper is central to the new work presented.

A.2.3 Keywords

Keywords are sometimes used for indexing. The keywords should not be words which appear in a title, but instead be fundamentally different ways of describing the content of the paper. For inter-disciplinary work, it is worth considering if the keywords would be understood by the different audience of another discipline, or if they are so technical that they can only be understood by one discipline.

A.2.4 Motivation

An instrument paper normally begins with some motivation for the work which was undertaken, which is usually about the reasons for the measurements concerned. This need not be a long section, but usually a broad statement about the general subject area would be made first, followed by several further sentences refining the topic to the nature of the problem to be discussed. References to previous work will be needed. It will almost certainly not be possible to mention all the relevant references, especially in a short note, but if there is a key historical reference, or a relevant review paper, it should be included, together with a few recent references.

A.2.5 Description

The next part of the paper seeks to describe the instrument from several perspectives, with the intention of conveying the principles used and any particular novel approach or technique needed for it to work correctly. This can be achieved using cogent and succinct text, particularly if it refers to a block diagram, an annotated functional drawing or an electronic circuit diagram. (Links in the text to the relevant figure being described will be needed.) In writing this, the object is not to provide an entire set of assembly instructions, but to provide sufficient information either to reproduce the instrument in another well-equipped development laboratory or for an instrumentation scientist to understand how the instrument functions. Annotated photographs may not copy very well when the work is reproduced, so schematic or outline diagrams are preferable in ensuring the work is understood correctly.

A.2.6 Comparison

The major finding of an instrumentation paper is usually about how well the instrument works and its range of application. This can most simply be achieved by a comparison with a better known or standard instrument, but alternatively it may require the generation of reference signals for calibration. In either case, a plot showing the instrument's response compared with either the standard instrument or standard values will be needed.

A.2.7 Figures

Figures should be compact, in that they should concentrate on displaying the available information obtained from the experiments. For this reason, extensive legends are worth avoiding, as they consume valuable plotting area of data. An explanatory caption is a useful alternative.

Figure captions themselves also require careful attention. A good caption is entirely self-contained, and provides all relevant information required to understand the figure without recourse to the main text. If the figure has several parts, the caption must address each of them.[i] Occasionally, it may be possible to provide a point or two of experimental detail in a caption, such as '… obtained using a hand-held anemometer under steady wind conditions', but the caption is not a place where the data should be interpreted in anything other than the simplest way (e.g. '… points concerned with the region of interest are marked'). Instead, the detailed discussion and interpretation should be provided in the main text, with a cross-link made to the figure when needed.

The convention of the x-axis carrying the independent variable (the quantity which can be changed) and the y-axis carrying the dependent variable (the response in the instrument), is usually interpreted to show values from the standard instrument on the x-axis with values from the instrument investigated shown on the y-axis. The scatter in the points can be informative, as are any error bars which can be added to the points. In some cases, a statistical fit determining the sensitivity of the response or a line allowing a comparison with theory may be a worthwhile addition.

Care is particularly needed in ensuring that the scales and units of the axes are clear and as simple as possible, for example using standard form or SI prefixes such

---

[i] A good test for whether the different parts of the figure can usefully be grouped together is to consider whether the combined caption is shorter than a set of individual captions.

as 'air temperature (°C)', 'droplet size (μm)' or 'thermistor resistance ($\times 10^4$ Ω)' to indicate that the quantities on the axis are in units of 10^4 Ω. If several plots are needed, these should be organised so that, as far as possible, axes with the same quantities are aligned, and have the same plotting range. For example, if two thermometers were separately compared with a standard thermometer, using two square plots aligned as a pair left and right with the same vertical axis range in each case would also allow a comparison between the two experiments. A further possibility in characterising the response is to plot the difference between the two instrument readings against the standard values.

A.2.8 Summary

The closing part of the paper collects together the findings, and emphasises the main points. There is no need to repeat what has been done except in the barest outline if there is a specific aspect to be drawn out, and there should be no repetition of earlier text or phrases in the Abstract. This is also not a place where new material is introduced. (If it is found that a fundamentally new point has to be made, an earlier section, such as the motivation or instrument comparison, should be modified and extended accordingly.)

A.2.9 Acknowledgements

Acknowledgements at the end should thank those on whose contributions the work's existence has depended, such as technicians and funding bodies, with a short grant reference code if appropriate.

A.3 Submission and revisions

The completed paper, checked for errors and correct figures, will be submitted to the selected instrumentation journal through its website. As for the peer-reviewed scientific literature in general, the journal will send a submitted paper to referees selected for the relevant expertise to the work discussed. After the referees have produced review reports, usually anonymously, the reviews will be sent to the author by the journal editor. The editor of the journal will usually indicate at this stage whether, if the review comments are addressed, the paper is likely to be published. (The editor may of course also decide to reject the paper on the basis of the review reports.)

In responding to review comments, it is conventional to deal with the points made separately, and indicate where in the manuscript changes have been made to address the review comments. If this is done clearly and straightforwardly, for example by marking the changes made in colour, it will make the job of the editor easier. Polite and constructive responses at this stage also form part of the orthodoxy of scientific interaction, in part because they may also offer the best prospects of conveying clarity of thought.

B

Anemometer Coordinate Rotations

Wind speed measurements made by a sonic anemometer along its particular measurement axes will, for research in micrometeorology and turbulence, generally need to be transformed for the flow orientation encountered [see also 128]. The rotation method to obtain measurements aligned with the flow can be represented by a series of matrix transformations, applied to the three components of raw wind instantaneous speed measurements (u_0, v_0, w_0) obtained on one set of physical directions in space. Here u_0 and v_0 are the horizontal wind components, and w_0 is vertically upwards. Associated with each of these directions are the mean wind speeds $\overline{u}_0$, $\overline{v}_0$ and $\overline{w}_0$ and their fluctuations about their mean values u'_0 v'_0 and w'_0. The mathematical process firstly establishes the horizontal plane, and then the vertical plane, and finally ensures that the momentum transfer is correctly aligned (see Figure B1). The coordinate rotation is performed for each averaging interval (typically 30 min).

(i) *Step 1: Rotation about the vertical axis*
This aligns the instantaneous wind direction in the mean horizontal plane of u_0 and v_0. The transformed instantaneous wind measurements u_1, v_1, and w_1 are

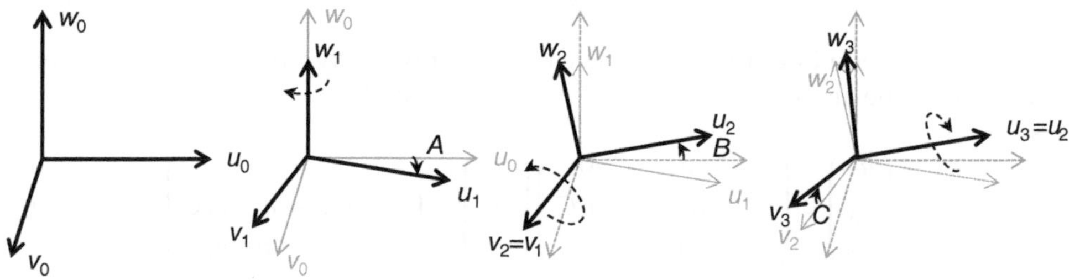

Figure B1 Coordinate rotations applied to orthogonal wind speed measurements (u_0, v_0, w_0). Firstly, these align in the horizontal plane through angle A along the mean wind direction (giving u_1, v_1, w_1), secondly a rotation by angle B about v_1 ensures the mean vertical wind speed is zero (giving u_2, v_2, w_2), and finally rotation by angle C ensures no lateral momentum transport (giving u_3, v_3, w_3).

Meteorological Measurements and Instrumentation, First Edition. R. Giles Harrison.
© 2015 John Wiley & Sons, Ltd. Published 2015 by John Wiley & Sons, Ltd.
Companion website: www.wiley.com/go/harrison/meteorologicalinstruments

found as

$$\begin{pmatrix} u_1 \\ v_1 \\ w_1 \end{pmatrix} = \begin{pmatrix} \cos A & \sin A & 0 \\ -\sin A & \cos A & 0 \\ 0 & 0 & 1 \end{pmatrix} \begin{pmatrix} u_0 \\ v_0 \\ w_0 \end{pmatrix} = \begin{pmatrix} u_0 \cos A + v_0 \sin A \\ v_0 \cos A - u_0 \sin A \\ w_0 \end{pmatrix}, \tag{B.1}$$

where

$$\tan A = \frac{\overline{v}_0}{\overline{u}_0}. \tag{B.2}$$

Note that the vertical wind speed component is unchanged ($w_1 = w_0$).

(ii) *Step 2: Rotation about the new horizontal axis*
This next transformation ensures that the mean velocity in the vertical direction perpendicular to the mean flow is zero, by solving for $\overline{w}_2 = 0$. The transformed instantaneous wind measurements u_2, v_2, and w_2 are found as

$$\begin{pmatrix} u_2 \\ v_2 \\ w_2 \end{pmatrix} = \begin{pmatrix} \cos B & 0 & \sin B \\ 0 & 1 & 0 \\ -\sin B & 0 & \cos B \end{pmatrix} \begin{pmatrix} u_1 \\ v_1 \\ w_1 \end{pmatrix} = \begin{pmatrix} u_1 \cos B + w_1 \sin B \\ v_1 \\ w_1 \cos B - u_1 \sin B \end{pmatrix}, \tag{B.3}$$

where the rotation angle B is found from $\overline{w_1 \cos B - u_1 \sin B} = 0$ as

$$\tan B = \frac{\overline{w}_0}{\sqrt{\overline{u}_0{}^2 + \overline{v}_0{}^2}}. \tag{B.4}$$

(iii) *Step 3: Final rotation*
This final step is not always implemented, but it ensures that the vertical direction is correctly aligned and that there is no lateral transport of momentum, that is, that

$$\overline{v_3' w_3'} = 0. \tag{B.5}$$

This is needed because, in some cases, the flow may not be genuinely horizontal, or the anemometer may not be perfectly vertical.

$$\begin{pmatrix} u_3 \\ v_3 \\ w_3 \end{pmatrix} = \begin{pmatrix} 1 & 0 & 0 \\ 0 & \cos C & \sin C \\ 0 & -\sin C & \cos C \end{pmatrix} \begin{pmatrix} u_2 \\ v_2 \\ w_2 \end{pmatrix} = \begin{pmatrix} u_2 \\ v_2 \cos C + w_2 \sin C \\ w_2 \cos C - v_2 \sin C \end{pmatrix}. \tag{B.6}$$

The rotation angle C is obtained by solving

$$\begin{aligned} \overline{v_3' w_3'} &= \overline{(v_2' \cos C + w_2' \sin C)(w_2' \cos C - v_2' \sin C)} \\ &= \overline{\left(w_2'^2 - v_2'^2\right)} \sin C \cos C + \overline{(w_2' v_2')}(\cos^2 C - \sin^2 C) \\ &= 0. \end{aligned} \tag{B.7}$$

This gives

$$\frac{\overline{v_2'w_2'}}{\overline{v_2'^2} - \overline{w_2'^2}} = \frac{\sin C \cos C}{\cos^2 C - \sin^2 C} = \frac{\frac{1}{2}\sin 2C}{\cos 2C} = \frac{1}{2}\tan 2C\,, \tag{B.8}$$

and hence the required angle C is

$$C = \frac{1}{2}\tan^{-1}\left[\frac{2\overline{v_2'w_2'}}{\overline{v_2'^2} - \overline{w_2'^2}}\right]. \tag{B.9}$$

References

1. L. Helmes and R. Jaenicke. 1985. Hidden information within series of measurements—four examples from atmospheric science. *J Atmos Chem*, 3, 171–185.
2. S. Emeis. 2000. Who created Réamur's thermometer scale? *Meteorologische Zeitschrift*, 9, 3, 185–187.
3. R.G. Harrison. 2005. Aurora diaries. *Astron & Geophys*, 46, 4.31–4.34.
4. G. Manley. 1974. Central England Temperatures: monthly means 1659 to 1973. *Quart Jour Roy Meteor Soc*, 100, 389–405.
5. D.E. Parker, T.P. Legg and C.K. Folland. 1992. A new daily Central England Temperature Series, 1772–1991. *Int J Clim*, 12, 317–342.
6. S.D. Burt. 2010. British Rainfall 1860–1993. *Weather*, 65, 121–128. doi: 10.1002/wea.603.
7. L.V. Alexander and P.D. Jones. 2001. Updated precipitation series for the U.K. and discussion of recent extremes. *Atmos Sci Lett*. doi:10.1006/asle.2001.0025.
8. W.E.K. Middleton, 1966. *A History of the Thermometer and its Uses in Meteorology*, The Johns Hopkins Press.
9. J. Glaisher. 1868. Description of thermometer stand. *Symons's Meteorol Mag*, 3, 155.
10. J. Aitken. 1884. On thermometer screens I. *Proc Roy Soc Edinburgh*, 12, 661–676; J. Aitken. 1884. On thermometer screens II. *Proc Roy Soc Edinburgh*, 12, 676–696; J. Aitken. 1886. On thermometer screens III. *Proc Roy Soc Edinburgh*, 13, 632–642; J. Aitken. 1887. On thermometer screens IV. *Proc Roy Soc Edinburgh*, 14, 428–432.
11. J. Aitken. 1921. Thermometer screens. *Proc Roy Soc Edinburgh*, 40, 172–181.
12. J. Laing. 1977. Maximum summer temperatures recorded in Glaisher stands and Stevenson screens. *Meteorol Mag*, 106, 220–228.
13. I.D. Margary. 1924. Glaisher stand versus Stevenson screen. A comparison of forty years' observations of maximum and minimum temperatures recorded in both screens at Camden Square, London. *Quart Jour Roy Meteor Soc*, 50, 209–223.
14. D.E. Parker. 1994. Effects of changing exposure of thermometers at land stations. *Int J Clim*, 14, 1–31.
15. E.G. Bilham. 1937. A screen for sheathed thermometers. *Quart Jour Roy Meteor Soc*, 63, 309–319.
16. G. Pfoetzner. 1972. History of the use of balloons in scientific experiments. *Space Sci Rev*, 13, 199–242.
17. R.G. Harrison and A.J. Bennett. 2007. Cosmic ray and air conductivity profiles retrieved from early twentieth century balloon soundings of the lower troposphere. *J Atmos Sol-Terr Phys*, 69, 4–5, 515–527. doi:10.1016/j.jastp.2006.09.008.
18. W.N. Shaw and W.H. Dines. 1904. Meteorological observations obtained by the use of kites off the west coast of Scotland, 1902. *Phil Trans A*, 202, 123–141.
19. K.C. Heidorn, The Advent of Regular Kite Observations. http://www.islandnet.com/~see/weather/almanac/arc2006/alm06apr2.htm (accessed 27 May 2014).

Meteorological Measurements and Instrumentation, First Edition. R. Giles Harrison.
© 2015 John Wiley & Sons, Ltd. Published 2015 by John Wiley & Sons, Ltd.
Companion website: www.wiley.com/go/harrison/meteorologicalinstruments

20. W.J. Humphreys. 1934. Radiometeorography as applied to unmanned balloons. *Month Weath Rev*, 62, 7, 221–226.
21. B.J. Booth. 2009. G.M.B. Dobson during world war 1 – his barothermograph and "bomb". *Weather*, 64, 8, 212–219.
22. R. Bureau. 1929. Sondages de pression et de temperature par radiotélégraphie. *Comptes Rendues*, 188, 1565–1566.
23. P. Idrac and R. Bureau. 1927. Expériences sur la propagation des ondes radiotélégraphiques en altitude. *Comptes Rendues*, 184, 691–692.
24. V. Väisälä. 1932. Bestrebungen und vorschläge zur entwicklung der radiometeorgraphischen methoden. Societas Scientarium Fennica (Helsingfors). *Commentationes Physico-Mathematicae*, 6, 2.
25. M.H.P. Ambaum and R.G. Harrison. 2011. The chaos machine: analogue computing rediscovered (1). *Elektor*, 416, 76–79; M.H.P. Ambaum, R.G. Harrison, J. Buiting and Th. Beckers. 2011. The chaos machine: analogue computing rediscovered (2). *Elektor*, 418, 72–75.
26. E.N. Lorenz. 1963. Deterministic nonperiodic flow. *J Atmos Sci*, 20(2), 130–141. doi:10.1175/1520-0469.
27. R.G. Harrison. 2002. A wide-range electrometer voltmeter for atmospheric measurements in thunderstorms and disturbed meteorological conditions. *Rev Sci Instrum*, 73(2), 482–483.
28. J.A. Chalmers. 1967. *Atmospheric Electricity*, 2nd edition, Pergamon Press.
29. R.G. Harrison. 1996. An atmospheric electrical voltmeter follower. *Rev Sci Instrum*, 67(7), 2636–2638.
30. R.G. Harrison and J.R. Knight. 2006. Thermopile radiometer signal conditioning for surface atmospheric radiation measurements. *Rev Sci Instrum*, 77, 116105. doi:10.1063/1.2370752.
31. K.A. Nicoll and R.G. Harrison. 2012. Balloon-borne disposable radiometer. *Rev Sci Instrum*, 83, 025111. doi:10.1063/1.3685252.
32. Y.B. Acharya and A.K. Aggarwal. 1996. Logarithmic current electrometer using light emitting diodes, *Meas Sci Technol*, 7, 151–156.
33. Y.B. Acharya and S.G. Tikekar. 1993. Low Current temperature compensated bipolar log-ratio amplifier. *Rev Sci Instrum*, 64, 6, 1652–1654.
34. G. Marlton, R.G. Harrison and K.A. Nicoll. 2013. Atmospheric point discharge current measurements using a temperature-compensated logarithmic current amplifier. *Rev Sci Instrum*, 84, 066103. doi:10.1063/1.4810849.
35. R.G. Harrison and K.L. Aplin. 2000. Femtoampere current reference stable over atmospheric temperatures *Rev Sci Instrum*, 71 (8), 3231–3232.
36. K.A. Nicoll. 2013. A self-calibrating electrometer for atmospheric charge measurements from a balloon platform. *Rev Sci Instrum*, 84, 096107.
37. R.G. Harrison and M.A. Pedder. 2001. Fine wire thermometer for air temperature measurement. *Rev Sci Instrum*, 72 (2), 1539–1541.
38. K.L. Aplin and R.G. Harrison. 2013. Lord Kelvin's atmospheric electricity measurements *History of Geo- and Space Sciences*, 4, 83–95. doi:10.5194/hgss-4-83-2013.
39. J.P. Cowley. 1976. Variations in global solar radiation at Kew. *Meteorol Mag*, 105(1252), 329–343.
40. L. Eyer and P. Bartholdi. 1999. Variable stars: which Nyquist frequency? *Astron Astrophys Supp Ser*, 135, 1.
41. R.G. Harrison, K.A. Nicoll and A.G. Lomas. 2012. Programmable data acquisition system for research measurements from meteorological radiosondes. *Rev Sci Instrum*, 83, 036106. doi:10.1063/1.3697717.
42. R.G. Harrison and B.N. Lodge. 1998. A calorimeter to detect freezing in supercooled water droplets. *Rev Sci Instrum*, 69, 11, 4004–4005.

43. M. Fulchignoni, A. Aboudan, F. Angrilli, M. Antonello, S. Bastianello, C. Bettanini, G. Bianchini, G. Colombatti, F. Ferri, E. Flamini, V. Gaborit, N. Ghafoor, B. Hathi, A.-M. Harri, A. Lehto, P.F. Lion Stoppato, M.R. Patel, J.C. Zarnecki. 2004. A stratospheric balloon experiment to test the Huygens atmospheric structure instrument (HASI). *Planetary and Space Science*, 52, 867–880. doi:10.1016/j.pss.2004.02.009.
44. D. Siomovitz and J. Joskowicz. 1980. Error evaluation of thermistor linearizing circuits. *Meas Sci Technol*, 1, 1280–1284.
45. R.G. Harrison and G.W. Rogers. 2006. Fine wire resistance thermometer amplifier for atmospheric measurements. *Rev Sci Instrum*, 77, 116112. doi:10.1063/1.2400013.
46. W.R. Sparks. 1972. The effect of thermometer screen design on the observed temperature. Technical Note 315, World Meteorological Organisation, Geneva.
47. E. Erell, V. Leal and E. Madonado. 2005. Measurement of air temperature in the presence of a large radiant flux: an assessment of passively ventilated thermometer screens. *Boundary Layer Meteorol*, 114, 205–231.
48. H.E. Painter. 1977. An analysis of the differences between dry-bulb temperatures obtained from an aspirated psychrometer and those from a naturally ventilated large thermometer screen at Kew Observatory. Meteorological Office unpublished report (National Meteorological Library, Exeter).
49. T. Andersson and I. Mattison. 1991. A field test of thermometer screens. Swedish Meteorological and Hydrological Institute, RMK No.62.
50. R.G. Harrison. 2010. Natural ventilation effects on temperatures within Stevenson screens. *Quart Jour Roy Meteorol Soc*, 136, 646, 253–259. doi:10.1002/qj.537.
51. WMO Guide to Meteorological Instruments and methods of observation WMO-No. 8 (2008 edition, updated in 2010) http://www.wmo.int/pages/prog/www/IMOP/CIMO-Guide.html (accessed May 2014).
52. D. Bryant. 1968. An investigation into the response of thermometer screens – the effect of wind speed on the lag time. *Meteorol Mag*, 97, 183–186.
53. R.G. Harrison. 2011. Lag-time effects on a naturally ventilated large thermometer screen. *Quart Jour Roy Meteorol Soc*, 137, 655, 402–408. doi:10.1002/qj.745.
54. M. Keil. 1996. *Temperature measurement in Stevenson screens*. MSc dissertation, Department of Meteorology, University of Reading.
55. M.C. Perry, M.J. Prior and D.E. Parker. 2007. An assessment of the suitability of a plastic thermometer screen for climatic data collection. *Int J Clim*, 27, 267–276.
56. S.P. Anderson and M.F. Baumgartner. 1998. Radiative heating errors in naturally ventilated air temperature measurements made from buoys. *J Atmos Ocean Tech*, 15, 1, 157–173.
57. M.G. Lawrence. 2005. The relationship between relative humidity and the dewpoint temperature in moist air a simple conversion and applications. *Bull Amer Meteorol Soc*, 86(2), 225–233.
58. D. Bolton. 1980. The computation of equivalent potential temperature. *Mon Weather Rev*, 108, 1046–1053.
59. P.R. Lowe and J.M. Ficke. 1974. Technical Paper 4–74, Environmental Prediction Res. Facility, Naval Post Grad. School, Monterey, California.
60. C.G. Wade. 1994. An evaluation of problems affecting the measurement of low relative humidity on the United States radiosonde. *J. Atmos Ocean Tech*, 11, 697–700.
61. A.C. Wilson, T.H. Barnes, P.J. Seakins, T.G. Rolfe and E.J. Meyer. 1995. A low-cost, high-speed, near-infrared hygrometer. *Rev Sci Instrum*, 66, 5618.
62. A.L. Buck. 1976. The variable-path Lyman-alpha hygrometer and its operating characteristics. *Bull Amer Meteorol Soc*, 57, 1113–1118.
63. K. Bartholomew, A. Illingworth and J. Nicol. 2010. Using radar to measure humidity: an overview. *Weather*, 65, 12, 332–334, doi:10.1002/wea.658.
64. R.G. Harrison and C.R. Wood, 2012. Ventilation effects on humidity measurements in thermometer screens. *Quart Jour Roy Meteorol Soc*, 138, 665, 1140–1120, doi:10.1002/qj.985.

65. N.D.S. Huband, S.C. King, M.W. Huxley and D.R. Butler. 1984. The performance of a thermometer screen on an automatic weather station. *Agric & Forest Meteorol*, 33, 2–3; see also pp. 249–258.
66. Moisture control with Philips' humidity sensor, Philips Components, http://home.utad.pt/~digital2/Apoio/Software/Sensores/Sensor_Humidade.pdf (accessed June 2014).
67. H. Spencer-Gregory and E. Rourke. 1957. *Hygrometry*. Crosby Lockwood & Son, London.
68. Meteorological Office. 1980. Handbook of Meteorological Instruments: Measurement of Atmospheric Pressure, 2nd edition, vol. 1. Her Majesty's Stationery Office, London. ISBN 0114003165.
69. D.H. McIntosh. 1972. *Meteorological Glossary*, Met Office, HMSO.
70. R.G. Harrison. 2000. A temperature-compensated meteorological barometer. *Rev Sci Instrum*, 71 (4), 1909–1910.
71. H. P. Gunnlaugsson, C. Holstein-Rathlou, J. P. Merrison, S. Knak Jensen, C. F. Lange, S. E. Larsen, M. B. Madsen, P. Nørnberg, H. Bechtold, E. Hald, J. J. Iversen, P. Lange, F. Lykkegaard, F. Rander, M. Lemmon, N. Renno, P. Taylor and P. Smith. 2008. Telltale wind indicator for the Mars Phoenix lander. *J Geophys Res*, 113, E00A04. doi:10.1029/2007JE003008.
72. C. Miller, J. Holmes, D. Henderson, J. Ginger and M. Morrison. 2013. The response of the Dines anemometer to gusts and comparisons with cup anemometers. *J Atmos Oceanic Technol*, 30, 1320–1336, doi:10.1175/JTECH-D-12-00109.1.
73. G.C. Gill. 1973. The helicoid anemometer. *Atmosphere*, 11 (4), 145–155.
74. E.G. Hauptmann. 1968. A simple hot wire anemometer probe. *J Phys E: Sci Instrum*, 1, 874.
75. C.M. Harris. 1969. Effects of humidity on the velocity of sound in air. *J Acoust Soc America*, 49(3), 890–893.
76. R.G. Harrison. 2001. Ultrasonic detection of atmospheric humidity variations. *Rev Sci Instrum*, 72(3), 1910–1913.
77. S.L. Gray and R.G. Harrison. 2012. Diagnosing eclipse-induced wind changes. *Proc Roy Soc Lond A*, 468, 2143, 1839–1850, doi:10.1098/rspa.2012.0007.
78. K. Walesby and R.G. Harrison. 2010. A thermally-stable tension meter for atmospheric soundings using kites. *Rev Sci Instrum*, 81, 076104, doi:10.1063/1.3465560.
79. R.M. Milne. 1921. Note on the Equation of Time. *Math Gaz*, 10, 372–375.
80. B.G. Collins and E.W. Walton. 1967. Temperature compensation of the Moll-Gorczynksi Pyranometer. *Meteor Mag*, 96, 1141, 225–288.
81. M.D. Steven and M.H. Unsworth. 1980. Shade-ring correction for pyranometer measurements of diffuse solar radiation from cloudless skies. *Quart Jour Roy Meteorol Soc*, 106, 865–872.
82. A.J. Drummond. 1956. On the measurement of sky radiation. *Arch Met Geophys Bioklm*, B7, 413–436.
83. K.L. Aplin and R.G. Harrison. 2003. Meteorological effects of the eclipse of 11th August 1999 in cloudy and clear conditions. *Proc Roy Soc Lond A*, 459, 2030, 353–372. doi:10.1098/rspa.2002.1042.
84. H.E. Painter. 1981. The performance of a Campbell–Stokes recorder compared with simultaneous record of the normal incidence irradiance. *Meteorol Mag*, 110, 102–109.
85. M. Bider. 1958. Über die Genauigkeit der Registrierungen des Sonnenscheinautographen Campbell–Stokes. *Arch Meteorol Geophys Bioklimatol, Serie B*, 9, 199–230.
86. A.M. Horseman, T. Richardson, A.T. Boardman, W. Tych, R. Timmis and A.R. MacKenzie. 2013. Calibrated digital images of Campbell–Stokes recorder card archives for direct solar irradiance studies. *Atmos Meas Tech*, 6, 1371–1379. doi:10.5194/amt-6-1371-2013.
87. H.L. Wright. 1935. Met Office Professional Notes, vol. 5, Professional Note 68.
88. C.R. Wood and R.G. Harrison. 2011. Mining the geophysical archive: scorch marks from the sky. *Weather*, 66(2), 39–41. doi:10.1002/wea.657.

89. C.G. Roberts. 1997. Trialling of an inexpensive electronic sunshine sensor June 1995 to May 1996. *Weather*, 52(12), 371–377.
90. J. Baker and J.E. Thornes. 2006. Solar position within Monet's Houses of Parliament. *Proc Roy Soc Lond A 8*, 462, 2076, 3775–3788.
91. K.L. Aplin and P.D. Williams. 2011. Meteorological phenomena in Western classical orchestral music. *Weather*, 66, 300–306.
92. H. Koschmieder. 1926. Theorie der horizontalen Sichtweite. *Beitr Phys freien Atmos*, 12, 33–55.
93. Met Office. 1982. Measurements of visibility and cloud height. *Handbook of Meteorological Instruments*, vol. 7, HMSO, London.
94. H.A. Douglas and D. Offiler. 1978. The Mk3 cloud recorder – a report on some of the potential accuracy limitations of this instrument. *Meteorol Mag*, 107, 23–32.
95. A.J. Bennett and R.G. Harrison. 2013. Lightning-induced extensive charge sheets provide long range electrostatic thunderstorm detection. *Phys Rev Lett*, 111, 045003.
96. R.G. Harrison. 2011. Fair weather atmospheric electricity. *J Phys: Conf Ser*, 301, 012001, doi:10.1088/1742-6596/301/1/012001.
97. M.J. Brettle and J.F.P. Galvin. 2003. Radiosondes: Part 1 – The instrument. *Weather* 57, 336–341.
98. E.G. Dymond. 1947. The Kew radio sonde. *Proc Phys Soc*, 59, 645–666. doi:10.1088/0959-5309/59/4/313.
99. H. Richner, J. Joss and P. Ruppert. 1996. A water hypsometer utilizing high-precision thermocouples. *J Atmos Oceanic Technol*, 13, 175–182. doi: 10.1175/1520-0426(1996)013<0175:AWHUHP>2.0.CO;2.
100. P.M. Rowe, L.M. Miloshevich, D.D. Turner and V.P. Walden. 2008. Dry bias in Vaisala RS90 radiosonde humidity profiles over antarctica. *J Atmos Oceanic Technol*, 25, 1529–1541. doi:10.1175/2008JTECHA1009.1.
101. B. Sun, A. Reale, D.J. Seidel and D.C. Hunt. 2010. Comparing radiosonde and COSMIC atmospheric profile data to quantify differences among radiosonde types and the effects of imperfect collocation on comparison statistics. *J Geophys Res*, 115, D23104. doi:10.1029/2010JD014457.
102. S.C. Sherwood, J.R. Lanzante and C.L. Meyer. 2005. Radiosonde daytime biases and late-20th century warming. *Science*, 309, 5740, 1556–1559. doi:10.1126/science.1115640.
103. R.G. Harrison. 2005. Inexpensive multichannel digital data acquisition system for a meteorological radiosonde. *Rev Sci Instrum*, 76, 026103. doi:10.1063/1.1841971.
104. G.C. Simpson and G.D. Robinson. 1940. The distribution of electricity in thunderclouds II. *Proc Roy Soc Lond A*, 177, 281–329.
105. M. Stolzenburg and T.C. Marshall. 2008. Charge structure and dynamics in thunderstorms. *Space Sci Rev*, 137. doi:10.1007/s11214-008-9338-z.
106. O.C. Jones, R.S. Maddever and J.H. Sanders. 1959. Radiosonde measurement of vertical electrical field and polar conductivity. *J Sci Instrum*, 36, 24–28.
107. K.A. Nicoll and R.G. Harrison. 2010. Experimental determination of layer cloud edge charging from cosmic ray ionisation. *Geophys Res Lett*, 37, L13802. doi:10.1029/2010GL043605.
108. A.W. Brewer and J.R. Milford. 1960. The Oxford-Kew Ozone Sonde. *Proc Roy Soc Lond A*, 256, 1287, 470–495.
109. Y.I. Stozhkov, N.S. Svirzhevsky, G.A. Bazilevskaya, A.N. Kvashnin, V.S. Makhmutov and A.K. Svirzhevskaya. 2009. Long-term (50 years) measurements of cosmic ray fluxes in the atmosphere. *Adv Space Res*, 44, 1124–1137.
110. S.W. Li, Y.S. Li and K.C. Tsui. 2007. Radioactivity in the atmosphere over Hong Kong. *J Environ Radioact*, 94, 98–106.
111. R.G. Harrison, K.A. Nicoll and A.G. Lomas. 2013. Geiger tube coincidence counter for lower atmosphere radiosonde measurements. *Rev Sci Instrum*, 84, 076103. doi:10.1063/1.4815832.

112. R. Philipona, A. Kräuchi and E. Brocard. 2012. Solar and thermal radiation profiles and radiative forcing measured through the atmosphere. *Geophys Res Lett*, 39, L13806. doi:10.1029/2012GL052087.
113. R.G. Harrison and R.J. Hogan. 2006. In-situ atmospheric turbulence measurement using the terrestrial magnetic field – a compass for a radiosonde. *J Atmos Oceanic Tech*, 23 (3), 517–523.
114. R.G. Harrison, A.M. Heath, R.J. Hogan and G.W. Rogers. 2009. Comparison of balloon-carried atmospheric motion sensors with Doppler lidar turbulence measurements. *Rev Sci Instrum*, 80(2), 026108. doi:10.1063/1.3086432.
115. G.E. Hill and D.S. Woffinden. 1980. A balloon borne instrument for the measurement of vertical profiles of supercooled liquid water concentration. *J Appl Meteorol*, 19, 1285–1292.
116. J.M. Rosen and D.J. Hofmann. 1977. Balloonborne measurements of condensation nuclei. *J Appl Meteor*, 16, 56–62. doi:10.1175/1520-0450(1977)016<0056:BMOCN>2.0.CO;2.
117. R.G. Harrison, K.A. Nicoll, Z. Ulanowski, T.A. Mather. 2010. Self-charging of the Eyjafjallajökull volcanic ash plume. *Environ Res Lett*, 5, 024004.
118. J.E. Simpson, D.A. Mansfield and J.R. Milford. 1977. Inland penetration of sea-breeze fronts. *QJR Meteorol Soc*, 103, 47–76. doi:10.1002/qj.49710343504.
119. N. Wildmann, M. Mauz and J. Bange. 2013. Two fast temperature sensors for probing of the atmospheric boundary layer using small remotely piloted aircraft (RPA). *Atmos Meas Tech*, 6, 2101–2113. doi:10.5194/amt-6-2101-2013.
120. K.G.T. Hollands. 1985. A derivation of the diffuse fraction's dependence on the clearness index. *Solar Energy*, 35, 2, 131–136.
121. M.H.P Ambaum, 2010. Significance tests in climate science. *J Climate*, 23, 5927–5932. doi:10.1175/2010JCLI3746.1.
122. A. Dai and J. Wang. 1999. Diurnal and semidiurnal tides in global pressure fields. *J Atmos Sci*, 56, 3874–3891.
123. G.C. Simpson. 1918. The twelve hourly barometer oscillation. *Quart Jour Roy Meteorol Soc*, 44, 1.
124. R.G. Harrison. 2013. The Carnegie curve. *Surv Geophys*, 34, 2, 209–232. doi:10.1007/s10712-012-9210-2.
125. R.G. Harrison, M. Joshi and K. Pascoe. 2011. Inferring convective responses to El Niño with atmospheric electricity measurements at Shetland. *Environ Res Lett*, 6, 044028.
126. R.G. Harrison and K.A. Nicoll. 2008. Air-earth current density measurements at Lerwick; implications for seasonality in the global electric circuit. *Atmos Res*, 89, 1–2, 181–193. doi:10.1016/j.atmosres.2008.01.008.
127. A.N. Kolmogorov. 1941. The local structure of turbulence in incompressible viscous fluids at very large Reynolds numbers. *Dokl Akad Nauk SSSR*, 30, 301–305. (Reprinted in 1991: *Proc Roy Soc Lond A*, 434, 9–13.)
128. T. Foken. 2008. Micrometeorology, Springer.

Index

Meteorological Measurements and Instrumentation, First Edition. R. Giles Harrison.

© 2015 John Wiley & Sons, Ltd. Published 2015 by John Wiley & Sons, Ltd.

Companion website: www.wiley.com/go/harrison/meteorologicalinstruments